Engineering Fundamentals of
Internal Combustion Engine

Engineering Fundamentals of Internal Combustion Engine

Edited by
Brody Walker

Larsen & Keller
www.larsen-keller.com

Engineering Fundamentals of Internal Combustion Engine
Edited by Brody Walker
ISBN: 978-1-63549-155-5 (Hardback)

© 2017 Larsen & Keller

☰ Larsen & Keller

Published by Larsen and Keller Education,
5 Penn Plaza,
19th Floor,
New York, NY 10001, USA

Cataloging-in-Publication Data

Engineering fundamentals of internal combustion engine / edited by Brody Walker.
    p. cm.
Includes bibliographical references and index.
ISBN 978-1-63549-155-5
1. Internal combustion engines. 2. Internal combustion engines--Combustion.
3. Internal combustion engine--Ignition. 4. Combustion engineering. I. Walker, Brody.
TJ759 .E64 2017
621.43--dc23

The publisher's policy is to use permanent paper from mills that operate a sustainable forestry policy. Furthermore, the publisher ensures that the text paper and cover boards used have met acceptable environmental accreditation standards.

Printed and bound in the United States of America.

For more information regarding Larsen and Keller Education and its products, please visit the publisher's website www.larsen-keller.com

Table of Contents

Preface

This book elucidates the concepts and innovative models around prospective developments with respect to internal combustion engine. It talks in detail about the techniques and applications of this technology. Internal combustion engine is an engine that relies on fuel combustion within a chamber to produce energy. It is used in powered aircrafts, jet engines, turbo engines, helicopters, etc. Through this book, we attempt to further enlighten the readers about the fundamental concepts in this field. It is a valuable compilation of topics, ranging from the basic to the most complex theories and principles in this field. The topics covered in this extensive book deal with the core subjects of ICE. This textbook aims to serve as a resource guide for students and experts alike and contribute to the growth of the discipline.

A short introduction to every chapter is written below to provide an overview of the content of the book:

Chapter 1 - This chapter will provide an integrated understanding of the internal combustion engine. Internal combustion engines usually refer to engines in which combustion exerts direct force to other parts of the engine. Common forms of these engines are the four stroke and two stroke engines. This section is an overview of the subject matter incorporating all the major aspects of internal combustion engine; **Chapter 2 -** Internal combustion engine is best understood in confluence with the major topics listed in the following chapter. The types of internal combustion engine explained in this chapter are diesel engine, petrol engine, four-stroke engine, two-stroke engine, jet engine, etc. The chapter strategically encompasses and incorporates the major components and key concepts of the internal combustion engine, providing a complete understanding; **Chapter 3 -** Cylinder block, piston, combustion chamber, spark plug and camshaft are some of the components of internal combustion engine. Most of these are discussed in this chapter. The topics discussed in the chapter are of great importance to broaden the existing knowledge on the subject matter; **Chapter 4 -** An ignition system heats an electrode. The electrode is heated to a very high temperature, mainly to ignite a fuel air mixture in this process. Types of ignition systems that have been developed are the ignition magneto ignition system, laser ignition system and inductive discharge ignition. This text provides the reader with an integrated study of ignition system; **Chapter 5 -** This chapter elucidates the important concepts and principles of the internal combustion engine. The important topics explained in this chapter are forced induction, manifold vacuum, consumption map and brake specific fuel consumption. The section strategically encompasses all the important concepts and helps the reader develop a better understanding on the principles of the internal combustion engine; **Chapter 6 -** Various engineers and scientists have contributed in the development of the internal combustion engine. Samuel Brown patented the first internal combustion engine which was proceeded by the design put forward by Nikolaus Otto. The history explicated in this chapter is very essential, as it educates the reader about the progress of the internal combustion engine.

Finally, I would like to thank my fellow scholars who gave constructive feedback and my family members who supported me at every step.

Editor

Introduction to Internal Combustion Engine

This chapter will provide an integrated understanding of the internal combustion engine. Internal combustion engines usually refer to engines in which combustion exerts direct force to other parts of the engine. Common forms of these engines are the four stroke and two stroke engines. This section is an overview of the subject matter incorporating all the major aspects of internal combustion engine.

An internal combustion engine (ICE) is a heat engine where the combustion of a fuel occurs with an oxidizer (usually air) in a combustion chamber that is an integral part of the working fluid flow circuit. In an internal combustion engine the expansion of the high-temperature and high-pressure gases produced by combustion apply direct force to some component of the engine. The force is applied typically to pistons, turbine blades, rotor or a nozzle. This force moves the component over a distance, transforming chemical energy into useful mechanical energy.

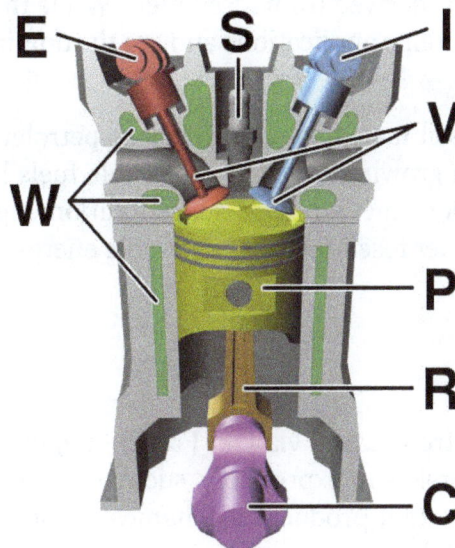

Diagram of a cylinder as found in 4-stroke gasoline engines.:C – crankshaft.E – exhaust camshaft.I – inlet camshaft.P – piston.R – connecting rod.S – spark plug.V – valves. red: exhaust, blue: intake.W – cooling water jacket. *gray structure* – engine block.

The first commercially successful internal combustion engine was created by Étienne Lenoir around 1859 and the first modern internal combustion engine was created in 1876 by Nikolaus Otto (see *Otto engine*).

The term *internal combustion engine* usually refers to an engine in which combustion is intermittent, such as the more familiar four-stroke and two-stroke piston engines, along with variants, such as the six-stroke piston engine and the Wankel rotary engine. A second class of internal combustion engines use continuous combustion: gas turbines, jet engines and most rocket engines, each of which are internal combustion engines on the same principle as previously described. Firearms are also a form of internal combustion engine.

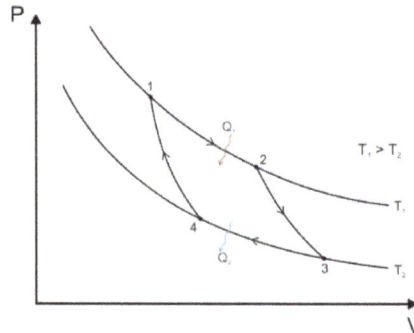

Diagram describing the ideal combustion cycle by Carnot

Internal combustion engines are quite different from external combustion engines, such as steam or Stirling engines, in which the energy is delivered to a working fluid not consisting of, mixed with, or contaminated by combustion products. Working fluids can be air, hot water, pressurized water or even liquid sodium, heated in a boiler. ICEs are usually powered by energy-dense fuels such as gasoline or diesel, liquids derived from fossil fuels. While there are many stationary applications, most ICEs are used in mobile applications and are the dominant power supply for vehicles such as cars, aircraft, and boats.

Typically an ICE is fed with fossil fuels like natural gas or petroleum products such as gasoline, diesel fuel or fuel oil. There's a growing usage of renewable fuels like biodiesel for compression ignition engines and bioethanol or methanol for spark ignition engines. Hydrogen is sometimes used, and can be made from either fossil fuels or renewable energy.

History

Etymology

At one time, the word *engine* (from Latin, via Old French, *ingenium*, "ability") meant any piece of machinery — a sense that persists in expressions such as *siege engine*. A "motor" (from Latin *motor*, "mover") is any machine that produces mechanical power. Traditionally, electric motors are not referred to as "Engines"; however, combustion engines are often referred to as "motors." (An *electric engine* refers to a locomotive operated by electricity.)

In boating an internal combustion engine that is installed in the hull is referred to as an engine, but the engines that sit on the transom are referred to as motors.

Applications

Reciprocating piston engines are by far the most common power source for land and water vehicles, including automobiles, motorcycles, ships and to a lesser extent, locomotives (some are electrical but most use Diesel engines). Rotary engines of the Wankel design are used in some automobiles, aircraft and motorcycles.

Reciprocating engine as found inside a car

Where very high power-to-weight ratios are required, internal combustion engines appear in the form of combustion turbines or Wankel engines. Powered aircraft typically uses an ICE which may be a reciprocating engine. Airplanes can instead use jet engines and helicopters can instead employ turboshafts; both of which are types of turbines. In addition to providing propulsion, airliners may employ a separate ICE as an auxiliary power unit. Wankel engines are fitted to many unmanned aerial vehicles.

Big Diesel generator used for backup power

Combined cycle power plant

ICEs drive some of the large electric generators that power electrical grids. They are found in the form of combustion turbines in combined cycle power plants with a typical electrical output in the range of 100 MW to 1 GW. The high temperature exhaust is used to boil and superheat water to run a steam turbine. Thus, the efficiency is higher because more energy is extracted from the fuel than what could be extracted by the combustion turbine alone. In combined cycle power plants efficiencies in the range of 50% to 60% are typical. In a smaller scale Diesel generators are used for backup power and for providing electrical power to areas not connected to an electric grid.

Small engines (usually 2-stroke gasoline engines) are a common power source for lawnmowers, string trimmers, chain saws, leafblowers, pressure washers, snowmobiles, jet skis, outboard motors, mopeds, and motorcycles.

Classification

There are several possible ways to classify internal combustion engines.

Reciprocating:

By number of strokes

- Two-stroke engine
- Clerk Cycle 1879
- Day Cycle
- Four-stroke engine (Otto cycle)
- Six-stroke engine

By type of ignition

- Compression-ignition engine
- Spark-ignition engine (commonly found as gasoline engines)

By mechanical/thermodynamical cycle (these 2 cycles do not encompass all reciprocating engines, and are infrequently used):

- Atkinson cycle
- Miller cycle

Rotary:

- Wankel engine

Continuous combustion:

- Gas turbine
- Jet engine
  - Rocket engine
  - Ramjet

The following jet engine types are also gas turbines types:

- Turbojet
- Turbofan
- Turboprop

Reciprocating Engines

Structure

Bare cylinder block of a V8 engine

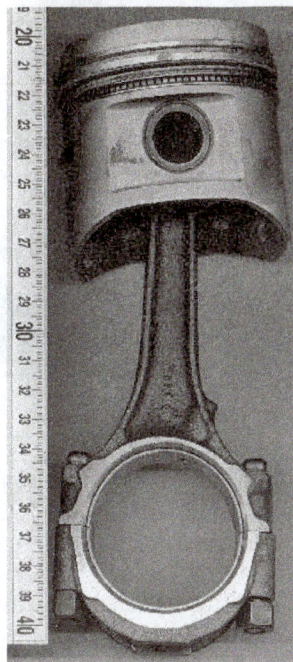

Piston, piston ring, gudgeon pin and connecting rod

The base of a reciprocating internal combustion engine is the engine block, which is typically made of cast iron or aluminium. The engine block contains the cylinders. In engines with more than one cylinder they are usually arranged either in 1 row (straight engine) or 2 rows (boxer engine or V engine); 3 rows are occasionally used (W engine) in contemporary engines, and other engine configurations are possible and have been used. Single cylinder engines are common for motorcycles and in small engines of machinery. Water-cooled engines contain passages in the engine block where cooling fluid circulates (the water jacket). Some small engines are air-cooled, and instead of having a water jacket the cylinder block has fins protruding away from it to cool by directly transferring heat to the air. The cylinder walls are usually finished by honing to obtain a cross hatch, which is better able to retain the oil. A too rough surface would quickly harm the engine by excessive wear on the piston.

The pistons are short cylindrical parts which seal one end of the cylinder from the high pressure of the compressed air and combustion products and slide continuously within it while the engine is in operation. The top wall of the piston is termed its *crown* and is typically flat or concave. Some two-stroke engines use pistons with a deflector head. Pistons are open at the bottom and hollow except for an integral reinforcement structure (the piston web). When an engine is working the gas pressure in the combustion chamber exerts a force on the piston crown which is transferred through its web to a gudgeon pin. Each piston has rings fitted around its circumference that mostly prevent the gases from leaking into the crankcase or the oil into the combustion chamber. A ventilation system drives the small amount of gas that escape past the pistons during normal operation (the blow-by gases) out of the crankcase so that it does not accumulate contaminating the oil and creating corrosion. In two-stroke gasoline engines the crankcase is part of the air–fuel path and due to the continuous flow of it they do not need a separate crankcase ventilation system.

Valve train above a Diesel engine cylinder head. This engine uses rocker arms but no pushrods.

The cylinder head is attached to the engine block by numerous bolts or studs. It has several functions. The cylinder head seals the cylinders on the side opposite to the pistons; it contains short ducts (the *ports*) for intake and exhaust and the associated intake valves that open to let the cylinder be filled with fresh air and exhaust valves that open to allow the combustion gases to escape. However, 2-stroke crankcase scavenged engines connect the gas ports directly to the cylinder wall without poppet valves; the piston controls their opening and occlusion instead. The cylinder head also holds the spark plug in the case of spark ignition engines and the injector for engines that use direct injection. All CI engines use fuel injection, usually direct injection but some engines instead use indirect injection. SI engines can use a carburetor or fuel injection as port injection or direct injection. Most SI engines have a single spark plug per cylinder but some have 2. A head gasket prevents the gas from leaking between the cylinder head and the engine block. The opening and closing of the valves is controlled by one or several camshafts and springs—or in some engines—a desmodromic mechanism that uses no springs. The camshaft may press directly the stem of the valve or may act upon a rocker arm, again, either directly or through a pushrod.

Engine block seen from below. The cylinders, oil spray nozzle and half of the main bearings are clearly visible.

The crankcase is sealed at the bottom with a sump that collects the falling oil during normal operation to be cycled again. The cavity created between the cylinder block and the sump houses a crankshaft that converts the reciprocating motion of the pistons to rotational motion. The crankshaft is held in place relative to the engine block by main bearings, which allow it to rotate. Bulkheads in the crankcase form a half of every main bearing; the other half is a detachable cap. In some cases a single *main bearing deck* is used rather than several smaller caps. A connecting rod is connected to offset sections of the crankshaft (the crankpins) in one end and to the piston in the other end through the gudgeon pin and thus transfers the force and translates the reciprocating motion of the pistons to the circular motion of the crankshaft. The end of the connecting rod attached to the gudgeon pin is called its small end, and the other end, where it is connected to the crankshaft, the big end. The big end has a detachable half to allow assembly around the crankshaft. It is kept together to the connecting rod by removable bolts.

The cylinder head has an intake manifold and an exhaust manifold attached to the corresponding ports. The intake manifold connects to the air filter directly, or to a carburetor when one is present, which is then connected to the air filter. It distributes the air incoming from these devices to the individual cylinders. The exhaust manifold is the first component in the exhaust system. It collects the exhaust gases from the cylinders and drives it to the following component in the path. The exhaust system of an ICE may also include a catalytic converter and muffler. The final section in the path of the exhaust gases is the tailpipe.

4-Stroke Engines

The *top dead center* (TDC) of a piston is the position where it is nearest to the valves; *bottom dead center* (BDC) is the opposite position where it is furthest from them. A *stroke* is the movement of a piston from TDC to BDC or vice versa together with the associated process. While an engine is in operation the crankshaft rotates continuously at a nearly constant speed. In a 4-stroke ICE each piston experiences 2 strokes per crankshaft revolution in the following order. Starting the description at TDC, these are:

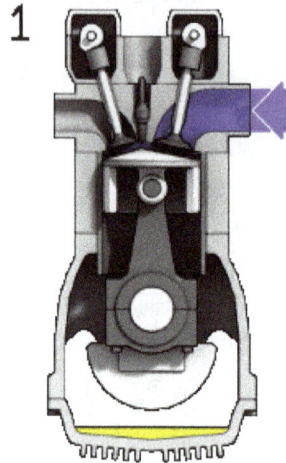

Diagram showing the operation of a 4-stroke SI engine. Labels:1 - Induction2 - Compression3 - Power4 - Exhaust

1. Intake, induction or suction: The intake valves are open as a result of the cam lobe pressing down on the valve stem. The piston moves downward increasing the volume of the combustion chamber and allowing air to enter in the case of a CI engine or an air fuel mix in the case of SI engines that do not use direct injection. The air or air-fuel mixture is called the *charge* in any case.

2. Compression: In this stroke, both valves are closed and the piston moves upward reducing the combustion chamber volume which reaches its minimum when the piston is at TDC. The piston performs work on the charge as it is being compressed; as a result its pressure, temperature and density increase; an approximation to this behavior is provided by the ideal gas law. Just before the piston reaches TDC, ignition begins. In the case of a SI engine, the spark plug receives a high voltage pulse that generates the spark which gives it its name and ignites the charge. In the case of a CI engine the fuel injector quickly injects fuel into the combustion chamber as a spray; the fuel ignites due to the high temperature.

3. Power or working stroke: The pressure of the combustion gases pushes the piston downward, generating more work than it required to compress the charge. Complementary to the compression stroke, the combustion gases expand and as a result their temperature, pressure and density decreases. When the piston is near to BDC the exhaust valve opens. The combustion gases expand irreversibly due to the leftover pressure—in excess of back pressure, the gauge pressure on the exhaust port—; this is called the *blowdown*.

4. Exhaust: The exhaust valve remains open while the piston moves upward expelling the combustion gases. For naturally aspirated engines a small part of the combustion gases may remain in the cylinder during normal operation because the piston does not close the combustion chamber completely; these gases dissolve in the next charge. At the end of this stroke, the exhaust valve closes, the intake valve opens, and the sequence repeats in the next cycle. The intake valve may open before the exhaust valve closes to allow better scavenging.

2-Stroke Engines

The defining characteristic of this kind of engine is that each piston completes a cycle every crankshaft revolution. The 4 processes of intake, compression, power and exhaust take place in only 2 strokes so that it is not possible to dedicate a stroke exclusively for each of them. Starting at TDC the cycle consist of:

1. Power: While the piston is descending the combustion gases perform work on it—as in a 4-stroke engine—. The same thermodynamic considerations about the expansion apply.

2. Scavenging: Around 75° of crankshaft rotation before BDC the exhaust valve or port opens, and blowdown occurs. Shortly thereafter the intake valve or transfer port opens. The incoming charge displaces the remaining combustion gases to the exhaust system and a part of the charge may enter the exhaust system as well. The piston reaches BDC and reverses direction. After the piston has traveled a short distance upwards into the cylinder the exhaust valve or port closes; shortly the intake valve or transfer port closes as well.

3. Compression: With both intake and exhaust closed the piston continues moving upwards compressing the charge and performing a work on it. As in the case of a 4-stroke engine, ignition starts just before the piston reaches TDC and the same consideration on the thermodynamics of the compression on the charge.

While a 4-stroke engine uses the piston as a positive displacement pump to accomplish scavenging taking 2 of the 4 strokes, a 2-stroke engine uses the last part of the power stroke and the first part of the compression stroke for combined intake and exhaust. The work required to displace the charge and exhaust gases comes from either the crankcase or a separate blower. For scavenging, expulsion of burned gas and entry of fresh mix, two main approaches are described: Loop scavenging, and Uniflow scavenging, SAE news published in the 2010s that 'Loop Scavenging' is better under any circumstance than Uniflow Scavenging.

Crankcase Scavenged

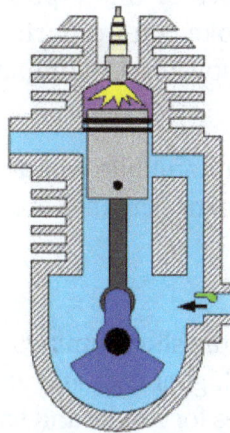

Diagram of a crankcase scavenged 2-stroke engine in operation

Some SI engines are crankcase scavenged and do not use poppet valves. Instead the crankcase and the part of the cylinder below the piston is used as a pump. The intake port is connected to the crankcase through a reed valve or a rotary disk valve driven by the engine. For each cylinder a transfer port connects in one end to the crankcase and in the other end to the cylinder wall. The exhaust port is connected directly to the cylinder wall. The transfer and exhaust port are opened and closed by the piston. The reed valve opens when the crankcase pressure is slightly below intake pressure, to let it be filled with a new charge; this happens when the piston is moving upwards. When the piston is moving downwards the pressure in the crankcase increases and the reed valve closes promptly, then the charge in the crankcase is compressed. When the piston is moving upwards, it uncovers the exhaust port and the transfer port and the higher pressure of the charge in the crankcase makes it enter the cylinder through the transfer port, blowing the exhaust gases. Lubrication is accomplished by adding *2-stroke oil* to the fuel in small ratios. *Petroil* refers to the mix of gasoline with the aforesaid oil. This kind of 2-stroke engines has a lower efficiency than comparable 4-strokes engines and release a more polluting exhaust gases for the following conditions:

- They use a *total-loss lubrication system*: all the lubricating oil is eventually burned along with the fuel.

- There are conflicting requirements for scavenging: On one side, enough fresh charge needs to be introduced in each cycle to displace almost all the combustion gases but introducing too much of it means that a part of it gets in the exhaust.

- They must use the transfer port(s) as a carefully designed and placed nozzle so that a gas current is created in a way that it sweeps the whole cylinder before reaching the exhaust port so as to expel the combustion gases, but minimize the amount of charge exhausted. 4-stroke engines have the benefit of forcibly expelling almost all of the combustion gases because during exhaust the combustion chamber is reduced to its minimum volume. In crankcase scavenged 2-stroke engines, exhaust and intake are performed mostly simultaneously and with the combustion chamber at its maximum volume.

The main advantage of 2-stroke engines of this type is mechanical simplicity and a higher power-to-weight ratio than their 4-stroke counterparts. Despite having twice as many power strokes per cycle, less than twice the power of a comparable 4-stroke engine is attainable in practice.

In the USA two stroke motorcycle and automobile engines were banned due to the pollution, although many thousands of lawn maintenance engines are in use.

Blower Scavenged

Using a separate blower avoids many of the shortcomings of crankcase scavenging, at the expense of increased complexity which means a higher cost and an increase in maintenance requirement. An engine of this type uses ports or valves for intake and valves for exhaust, except opposed piston engines, which may also use ports for exhaust. The blower is usually of the Roots-type but other types have been used too. This design is commonplace in CI engines, and has been occasionally used in SI engines.

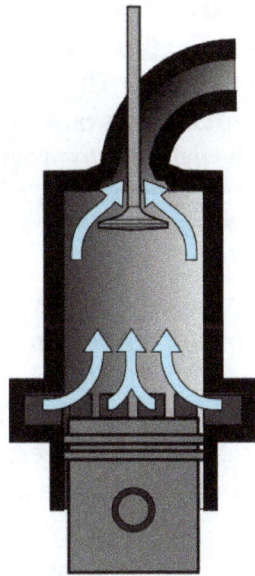

Diagram of uniflow scavenging

CI engines that use a blower typically use *uniflow scavenging*. In this design the cylinder wall contains several intake ports placed uniformly spaced along the circumference just above the position that the piston crown reaches when at BDC. An exhaust valve or several like that of 4-stroke engines is used. The final part of the intake manifold is an air sleeve which feeds the intake ports. The intake ports are placed at an horizontal angle to the cylinder wall (I.e: they are in plane of the piston crown) to give a swirl to the incoming charge to improve combustion. The largest reciprocating IC are low speed CI engines of this type; they are used for marine propulsion (see marine diesel engine) or electric power generation and achieve the highest thermal efficiencies among internal combustion engines of any kind. Some Diesel-electric locomotive engines operate on the 2-stroke cycle. The most powerful of them have a brake power of around 4.5 MW or 6,000 HP. The EMD SD90MAC class of locomotives use a 2-stroke engine. The comparable class GE AC6000CW whose prime mover has almost the same brake power uses a 4-stroke engine.

An example of this type of engine is the Wärtsilä-Sulzer RTA96-C turbocharged 2-stroke Diesel, used in large container ships. It is the most efficient and powerful internal combustion engine in the world with a thermal efficiency over 50%. For comparison, the most efficient small four-stroke engines are around 43% thermally-efficient (SAE 900648); size is an advantage for efficiency due to the increase in the ratio of volume to surface area.

See the external links for a in-cylinder combustion video in a 2-stroke, optically accessible motorcycle engine.

Historical Design

Dugald Clerk developed the first two cycle engine in 1879. It used a separate cylinder which functioned as a pump in order to transfer the fuel mixture to the cylinder.

In 1899 John Day simplified Clerk's design into the type of 2 cycle engine that is very widely used today. Day cycle engines are crankcase scavenged and port timed. The crankcase and the part of the cylinder below the exhaust port is used as a pump. The operation of the Day cycle engine begins when the crankshaft is turned so that the piston moves from BDC upward (toward the head) creating a vacuum in the crankcase/cylinder area. The carburetor then feeds the fuel mixture into the crankcase through a reed valve or a rotary disk valve (driven by the engine). There are cast in ducts from the crankcase to the port in the cylinder to provide for intake and another from the exhausst port to the exhaust pipe. The height of the port in relationship to the length of the cylinder is called the "port timing."

On the first upstroke of the engine there would be no fuel inducted into the cylinder as the crankcase was empty. On the downstroke the piston now compresses the fuel mix, which has lubricated the piston in the cylinder and the bearings due to the fuel mix having oil added to it. As the piston moves downward is first uncovers the exhaust, but on the first stroke there is no burnt fuel to exhaust. As the piston moves downward further, it uncovers the intake port which has a duct that runs to the crankcase. Since the fuel mix in the crankcase is under pressure the mix moves through the duct and into the cylinder.

Because there is no obstruction in the cylinder of the fuel to move directly out of the exhaust port prior to the piston rising far enough to close the port, early engines used a high domed piston to slow down the flow of fuel. Later the fuel was "resonated" back into the cylinder using an expansion chamber design. When the piston rose close to TDC a spark ignites the fuel. As the piston is driven downward with power it first uncovers the exhaust port where the burned fuel is expelled under high pressure and then the intake port where the process has been completed and will keep repeating.

Later engines used a type of porting devised by the Deutz company to improve performance. It was called the Schnurle Reverse Flow system. DKW licensed this design for all their motorcycles. Their DKW RT 125 was one of the first motor vehicles to achieve over 100 mpg as a result.

Ignition

Internal combustion engines require ignition of the mixture, either by spark ignition (SI) or compression ignition (CI). Before the invention of reliable electrical methods, hot tube and flame methods were used. Experimental engines with laser ignition have been built.

Spark Ignition Process

The spark ignition engine was a refinement of the early engines which used Hot Tube ignition. When Bosch developed the magneto it became the primary system for producing electricity to energize a spark plug. Many small engines still use magneto ignition. Small engines are started by hand cranking using a recoil starter or hand crank . Prior to Charles F. Kettering of Delco's development of the automotive starter all gasoline engined automobiles used a hand crank.

Larger engines typically power their starting motors and Ignition systems using using the electrical energy stored in a lead–acid battery. The battery's charged state is maintained by an automotive alternator or (previously) a generator which uses engine power to create electrical energy storage.

Bosch Magneto

Points and Coil Ignition

The battery supplies electrical power for starting when the engine has a starting motor system, and supplies electrical power when the engine is off. The battery also supplies electrical power during rare run conditions where the alternator cannot maintain more than 13.8 volts (for a common 12V automotive electrical system). As alternator voltage falls below 13.8 volts, the lead-acid storage battery increasingly picks up electrical load. During virtually all running conditions, including normal idle conditions, the alternator supplies primary electrical power.

Some systems disable alternator field (rotor) power during wide open throttle conditions. Disabling the field reduces alternator pulley mechanical loading to nearly zero, maximizing crankshaft power. In this case the battery supplies all primary electrical power.

Gasoline engines take in a mixture of air and gasoline and compress it by the movement of the piston from bottom dead center to top dead center when the fuel is at maximum compression. The reduction in the size of the swept area of the cylinder and taking into account the volume of the combustion chamber is described by a ratio. Early engines had compression ratios of 6 to 1. As compression ratios were increased the efficiency of the engine increased as well.

With early induction and ignition systems the compression ratios had to be kept low. With advances in fuel technology and combustion management high performance engines can run reliably at 12:1 ratio. With low octane fuel a problem would occur as the compression ratio increased as the fuel was igniting due to the rise in temperature that resulted. Charles Kettering developed a lead additive which allowed higher compression ratios.

The fuel mixture is ignited at difference progressions of the piston in the cylinder. At low rpm the spark is timed to occur close to the piston achieving top dead center. In order to produce more power, as rpm rises the spark is advanced sooner during piston movement. The spark occurs while the fuel is still being compressed progressively more as rpm rises.

The necessary high voltage, typically 10,000 volts, is supplied by an induction coil or transformer. The induction coil is a fly-back system, using interruption of electrical primary system current through some type of synchronized interrupter. The interrupter can be either contact points or a power transistor. The problem with this type of ignition is that as RPM increases the available of electrical energy decreases. This is especially as problem since the amount of energy needed to ignite a more dense fuel mixture is higher. The result was often a high rpm misfire.

Capacitor discharge ignition was developed. It produces a rising voltage that is sent to the spark plug. CD system voltages can reach 60,000 volts. CD ignitions use step-up transformers. The step-up transformer uses energy stored in a capacitance to generate electric spark. With either system, a mechanical or electrical control system provides a carefully timed high-voltage to the proper cylinder. This spark, via the spark plug, ignites the air-fuel mixture in the engine's cylinders.

While gasoline internal combustion engines are much easier to start in cold weather than diesel engines, they can still have cold weather starting problems under extreme conditions. For years the solution was to park the car in heated areas. In some parts of the world the oil was actually drained and heated over night and returned to the engine for cold starts. In the early 1950s the gasoline Gasifier unit was developed, where, on cold weather starts, raw gasoline was diverted to the unit where part of the fuel was burned causing the other part to become a hot vapor sent directly to the intake valve manifold. This unit was quite popular until electric engine block heaters became standard on gasoline engines sold in cold climates.

Diesel Ignition Process

Diesel engines and HCCI (Homogeneous charge compression ignition) engines, rely solely on heat and pressure created by the engine in its compression process for ignition. The compression level that occurs is usually twice or more than a gasoline engine. Diesel engines take in air only, and shortly before peak compression, spray a small quantity of diesel fuel into the cylinder via a fuel injector that allows the fuel to instantly ignite. HCCI type engines take in both air and fuel, but continue to rely on an unaided auto-combustion process, due to higher pressures and heat. This is also why diesel and HCCI engines are more susceptible to cold-starting issues, although they run just as well in cold weather once started. Light duty diesel engines with indirect injection in automobiles and light trucks employ glowplugs (or other pre-heating: see Cummins ISB#6BT) that pre-heat the combustion chamber just before starting to reduce no-start conditions in cold weather. Most diesels also have a battery and charging system; nevertheless, this system is secondary and is added by manufacturers as a luxury for the ease of starting, turning fuel on and off (which can also be done via a switch or mechanical apparatus), and for running auxiliary electrical components and accessories. Most new engines rely on electrical and electronic engine control units (ECU) that also adjust the combustion process to increase efficiency and reduce emissions.

Lubrication

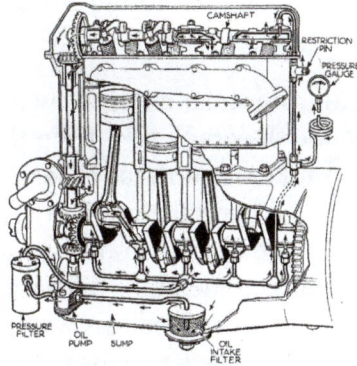

Diagram of an engine using pressurized lubrication

Surfaces in contact and relative motion to other surfaces require lubrication to reduce wear, noise and increase efficiency by reducing the power wasting in overcoming friction, or to make the mechanism work at all. At the very least, an engine requires lubrication in the following parts:

- Between pistons and cylinders

- Small bearings

- Big end bearings

- Main bearings

- Valve gear (The following elements may not be present):

    o Tappets

    o Rocker arms

    o Pushrods

    o Timing chain or gears. Toothed belts do not require lubrication.

In 2-stroke crankcase scavenged engines, the interior of the crankcase, and therefore the crankshaft, connecting rod and bottom of the pistons are sprayed by the 2-stroke oil in the air-fuel-oil mixture which is then burned along with the fuel. The valve train may be contained in a compartment flooded with lubricant so that no oil pump is required.

In a *splash lubrication system* no oil pump is used. Instead the crankshaft dips into the oil in the sump and due to its high speed, it splashes the crankshaft, connecting rods and bottom of the pistons. The connecting rod big end caps may have an attached scoop to enhance this effect. The valve train may also be sealed in a flooded compartment, or open to the crankshaft in a way that it receives splashed oil and allows it to drain back to the sump. Splash lubrication is common for small 4-stroke engines.

In a *forced* (also called *pressurized*) *lubrication system*, lubrication is accomplished in a closed loop which carries motor oil to the surfaces serviced by the system and then returns the oil to a reservoir. The auxiliary equipment of an engine is typically not serviced by this loop; for instance, an alternator may use ball bearings sealed with its lubricant. The reservoir for the oil is usually the sump, and when this is the case, it is called a *wet sump* system. When there is a different oil reservoir the crankcase still catches it, but it is continuously drained by a dedicated pump; this is called a *dry sump* system.

On its bottom, the sump contains an oil intake covered by a mesh filter which is connected to an oil pump then to an oil filter outside the crankcase, from there it is diverted to the crankshaft main bearings and valve train. The crankcase contains at least one *oil gallery* (a conduit inside a crankcase wall) to which oil is introduced from the oil filter. The main bearings contain a groove through all or half its circumference; the oil enters to these grooves from channels connected to the oil gallery. The crankshaft has drillings which take oil from these grooves and deliver it to the big end bearings. All big end bearings are lubricated this way. A single main bearing may provide oil for 0, 1 or 2 big end bearings. A similar system may be used to lubricate the piston, its gudgeon pin and the small end of its connecting rod; in this system, the connecting rod big end has a groove around the crankshaft and a drilling connected to the groove which distributes oil from there to the bottom of the piston and from then to the cylinder.

Other systems are also used to lubricate the cylinder and piston. The connecting rod may have a nozzle to throw an oil jet to the cylinder and bottom of the piston. That nozzle is in movement relative to the cylinder it lubricates, but always pointed towards it or the corresponding piston.

Typically a forced lubrication systems have a lubricant flow higher than what is required to lubricate satisfactorily, in order to assist with cooling. Specifically, the lubricant system helps to move heat from the hot engine parts to the cooling liquid (in water-cooled engines) or fins (in air-cooled engines) which then transfer it to the environment. The lubricant must be designed to be chemically stable and maintain suitable viscosities within the temperature range it encounters in the engine.

Cylinder Configuration

Common cylinder configurations include the straight or inline configuration, the more compact V configuration, and the wider but smoother flat or boxer configuration. Aircraft engines can also adopt a radial configuration, which allows more effective cooling. More unusual configurations such as the H, U, X, and W have also been used.

Multiple cylinder engines have their valve train and crankshaft configured so that pistons are at different parts of their cycle. It is desirable to have the piston's cycles uniformly spaced (this is called *even firing*) especially in forced induction engines; this reduces torque pulsations and makes inline engines with more than 3 cylinders statically balanced in its primary forces. However, some engine configurations require odd firing to achieve better balance than what is possible with even firing. For instance, a 4-stroke I2 engine has better balance when the angle between the crankpins is 180° because the pistons move in opposite directions and inertial forces partially cancel, but this gives an odd firing pattern where one cylinder fires 180° of crankshaft rotation after the other, then no cylinder fires for 540°. With an even firing pattern the pistons would move in unison and the associated forces would add.

Multiple crankshaft configurations do not necessarily need a cylinder head at all because they can instead have a piston at each end of the cylinder called an opposed piston design. Because fuel inlets and outlets are positioned at opposed ends of the cylinder, one can achieve uniflow scavenging, which, as in the four-stroke engine is efficient over a wide range of engine speeds. Thermal efficiency is improved because of a lack of cylinder heads. This design was used in the Junkers Jumo 205 diesel aircraft engine, using two crankshafts at either end of a single bank of cylinders, and most remarkably in the Napier Deltic diesel engines. These used three crankshafts to serve three banks of double-ended cylinders arranged in an equilateral triangle with the crankshafts at the corners. It was also used in single-bank locomotive engines, and is still used in marine propulsion engines and marine auxiliary generators.

Diesel Cycle

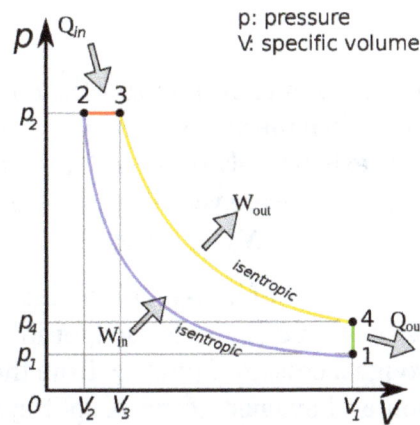

P-v Diagram for the Ideal Diesel cycle. The cycle follows the numbers 1–4 in clockwise direction.

Most truck and automotive diesel engines use a cycle reminiscent of a four-stroke cycle, but with a compression heating ignition system, rather than needing a separate ignition system. This variation is called the diesel cycle. In the diesel cycle, diesel fuel is injected directly into the cylinder so that combustion occurs at constant pressure, as the piston moves.

Otto cycle: Otto cycle is the typical cycle for most of the cars internal combustion engines, that work using gasoline as a fuel. Otto cycle is exactly the same one that was described for the four-stroke engine. It consists of the same four major steps: Intake, compression, ignition and exhaust.

PV diagram for Otto cycle On the PV-diagram, 1–2: Intake: suction stroke 2–3: Isentropic Compression stroke 3–4: Heat addition stroke 4–5: Exhaust stroke (Isentropic expansion) 5–2: Heat rejection The distance between points 1–2 is the stroke of the engine. By dividing V_2/V_1, we get: r, where r is called the compression ratio of the engine.

Five-stroke Engine

In 1879, Nikolaus Otto manufactured and sold a double expansion engine (the double and triple expansion principles had ample usage in steam engines), with two small cylinders at both sides of a low-pressure larger cylinder, where a second expansion of exhaust stroke gas took place; the owner returned it, alleging poor performance. In 1906, the concept was incorporated in a car built by EHV (Eisenhuth Horseless Vehicle Company) CT, USA; and in the 21st century Ilmor designed and successfully tested a 5-stroke double expansion internal combustion engine, with high power output and low SFC (Specific Fuel Consumption).

Six-stroke Engine

The six-stroke engine was invented in 1883. Four kinds of six-stroke use a regular piston in a regular cylinder (Griffin six-stroke, Bajulaz six-stroke, Velozeta six-stroke and Crower six-stroke), firing every three crankshaft revolutions. The systems capture the wasted heat of the four-stroke Otto cycle with an injection of air or water.

The Beare Head and "piston charger" engines operate as opposed-piston engines, two pistons in a single cylinder, firing every two revolutions rather more like a regular four-stroke.

Other Cycles

The very first internal combustion engines did not compress the mixture. The first part of the piston downstroke drew in a fuel-air mixture, then the inlet valve closed and, in the remainder of the downstroke, the fuel-air mixture fired. The exhaust valve opened for the piston upstroke. These attempts at imitating the principle of a steam engine were very inefficient. There are a number of variations of these cycles, most notably the Atkinson and Miller cycles. The diesel cycle is somewhat different.

Split-cycle engines separate the four strokes of intake, compression, combustion and exhaust into two separate but paired cylinders. The first cylinder is used for intake and compression. The compressed air is then transferred through a crossover passage from the compression cylinder into the second cylinder, where combustion and exhaust occur. A split-cycle engine is really an air compressor on one side with a combustion chamber on the other.

Previous split-cycle engines have had two major problems—poor breathing (volumetric efficiency) and low thermal efficiency. However, new designs are being introduced that seek to address these problems.

The Scuderi Engine addresses the breathing problem by reducing the clearance between the piston and the cylinder head through various turbo charging techniques. The Scuderi design requires the use of outwardly opening valves that enable the piston to move very close to the cylinder head without the interference of the valves. Scuderi addresses the low thermal efficiency via firing after top dead centre (ATDC).

Firing ATDC can be accomplished by using high-pressure air in the transfer passage to create sonic flow and high turbulence in the power cylinder.

Combustion Turbines

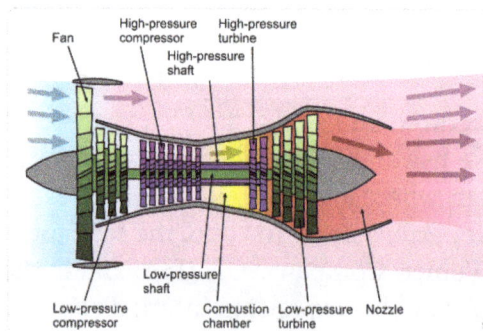

Turbofan Jet Engine

Jet engines use a number of rows of fan blades to compress air which then enters a combustor where it is mixed with fuel (typically JP fuel) and then ignited. The burning of the fuel raises the temperature of the air which is then exhausted out of the engine creating thrust. A modern turbo-fan engine can operate at as high as 48% efficiency.

There are six sections to a Fan Jet engine:

- Fan

- Compressor

- Combustor

- Turbine

- Mixer

- Nozzle

Gas Turbines

Turbine Power Plant

A gas turbine compresses air and uses it to turn a turbine. It is essentially a Jet engine which directs it's output to a shaft. There are three stages to a turbine: 1) air is drawn through a compressor where the temperature rises due to compression, 2) fuel is added in the combuster, and 3) hot air is exhausted through turbines blades which rotate a shaft connected to the compressor.

A gas turbine is a rotary machine similar in principle to a steam turbine and it consists of three main components: a compressor, a combustion chamber, and a turbine. The air, after being compressed in the compressor, is heated by burning fuel in it. About ⅔ of the heated air, combined with the products of combustion, expands in a turbine, producing work output that drives the compressor. The rest (about ⅓) is available as useful work output.

Gas Turbines are among the MOST efficient internal combustion engines. The General Electric 7HA and 9HA turbine electrical plants are rated at over 61% efficiency.

Brayton Cycle

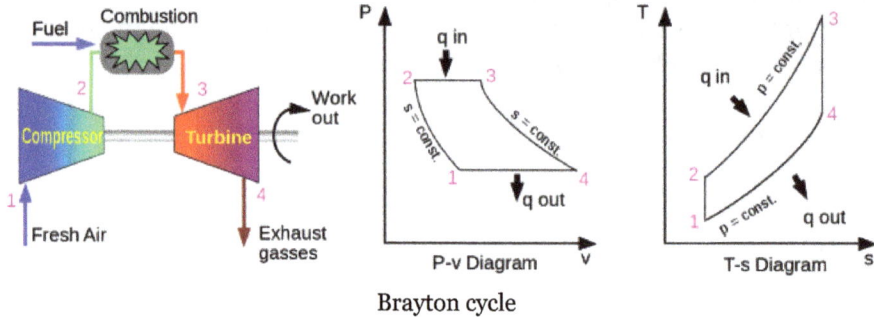

Brayton cycle

A gas turbine is a rotary machine somewhat similar in principle to a steam turbine. It consists of three main components: compressor, combustion chamber, and turbine. The air is compressed by the compressor where a temperature rise occurs. The compressed air is further heated by combustion of injected fuel in the combustion chamber which expands the air. This energy rotates the turbine which powers the compressor via a mechanical coupling. The hot gases are then exhausted to provide thrust.

Gas turbine cycle engines employ a continuous combustion system where compression, combustion, and expansion occur simultaneously at different places in the engine—giving continuous power. Notably, the combustion takes place at constant pressure, rather than with the Otto cycle, constant volume.

Wankel Engines

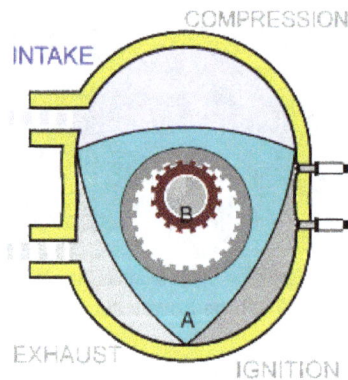

The Wankel rotary cycle. The shaft turns three times for each rotation of the rotor around the lobe and once for each orbital revolution around the eccentric shaft.

The Wankel engine (rotary engine) does not have piston strokes. It operates with the same separation of phases as the four-stroke engine with the phases taking place in separate locations in the engine. In thermodynamic terms it follows the Otto engine cycle, so may be thought of as a "four-phase" engine. While it is true that three power strokes typically occur per rotor revolution, due to the 3:1 revolution ratio of the rotor to the eccentric shaft, only one power stroke per shaft revolution actually occurs. The drive (eccentric) shaft rotates once during every power stroke instead of twice (crankshaft), as in the Otto cycle, giving it a greater power-to-weight ratio than piston engines. This type of engine was most notably used in the Mazda RX-8, the earlier RX-7, and other vehicle models. The engine is also use in unmanned aerial vehicles, where the small size and weight and the high power-to-weight ratio are advantages.

Forced Induction

Forced induction is the process of delivering compressed air to the intake of an internal combustion engine. A forced induction engine uses a gas compressor to increase the pressure, temperature and density of the air. An engine without forced induction is considered a naturally aspirated engine.

Forced induction is used in the automotive and aviation industry to increase engine power and efficiency. It particularly helps aviation engines, as they need to operate at high altitude.

Forced induction is achieved by a supercharger, where the compressor is directly powered from the engine shaft or, in the turbocharger, from a turbine powered by the engine exhaust.

Fuels and Oxidizers

All internal combustion engines depend on combustion of a chemical fuel, typically with oxygen from the air (though it is possible to inject nitrous oxide to do more of the same thing and gain a power boost). The combustion process typically results in the production of a great quantity of heat, as well as the production of steam and carbon dioxide and other chemicals at very high temperature; the temperature reached is determined by the chemical make up of the fuel and oxidisers (see stoichiometry), as well as by the compression and other factors.

Fuels

The most common modern fuels are made up of hydrocarbons and are derived mostly from fossil fuels (petroleum). Fossil fuels include diesel fuel, gasoline and petroleum gas, and the rarer use of propane. Except for the fuel delivery components, most internal combustion engines that are designed for gasoline use can run on natural gas or liquefied petroleum gases without major modifications. Large diesels can run with air mixed with gases and a pilot diesel fuel ignition injection. Liquid and gaseous biofuels, such as ethanol and biodiesel (a form of diesel fuel that is produced from crops that yield triglycerides such as soybean oil), can also be used. Engines with appropriate modifications can also run on hydrogen gas, wood gas, or charcoal gas, as well as from so-called producer gas made from other convenient biomass. Experiments have also been conducted using powdered solid fuels, such as the magnesium injection cycle.

Presently, fuels used include:

- Petroleum:

    o   Petroleum spirit (North American term: gasoline, British term: petrol)

    o   Petroleum diesel.

    o   Autogas (liquified petroleum gas).

    o   Compressed natural gas.

    o   Jet fuel (aviation fuel)

    o   Residual fuel

- Coal:

  o Gasoline can be made from carbon (coal) using the Fischer-Tropsch process

  o Diesel fuel can be made from carbon using the Fischer-Tropsch process

- Biofuels and vegetable oils:

  o Peanut oil and other vegetable oils.

  o Woodgas, from an onboard wood gasifier using solid wood as a fuel

  o Biofuels:

    ▪ Biobutanol (replaces gasoline).

    ▪ Biodiesel (replaces petrodiesel).

    ▪ Dimethyl Ether (replaces petrodiesel).

    ▪ Bioethanol and Biomethanol (wood alcohol) and other biofuels (see Flexible-fuel vehicle).

    ▪ Biogas

- Hydrogen (mainly spacecraft rocket engines)

Even fluidized metal powders and explosives have seen some use. Engines that use gases for fuel are called gas engines and those that use liquid hydrocarbons are called oil engines; however, gasoline engines are also often colloquially referred to as, "gas engines" ("petrol engines" outside North America).

The main limitations on fuels are that it must be easily transportable through the fuel system to the combustion chamber, and that the fuel releases sufficient energy in the form of heat upon combustion to make practical use of the engine.

Diesel engines are generally heavier, noisier, and more powerful at lower speeds than gasoline engines. They are also more fuel-efficient in most circumstances and are used in heavy road vehicles, some automobiles (increasingly so for their increased fuel efficiency over gasoline engines), ships, railway locomotives, and light aircraft. Gasoline engines are used in most other road vehicles including most cars, motorcycles, and mopeds. Note that in Europe, sophisticated diesel-engined cars have taken over about 45% of the market since the 1990s. There are also engines that run on hydrogen, methanol, ethanol, liquefied petroleum gas (LPG), biodiesel, paraffin and tractor vaporizing oil (TVO).

Hydrogen

Hydrogen could eventually replace conventional fossil fuels in traditional internal combustion engines. Alternatively fuel cell technology may come to deliver its promise and the use of the internal combustion engines could even be phased out.

Although there are multiple ways of producing free hydrogen, those methods require converting combustible molecules into hydrogen or consuming electric energy. Unless that electricity is produced from a renewable source—and is not required for other purposes— hydrogen does not solve any energy crisis. In many situations the disadvantage of hydrogen, relative to carbon fuels, is its storage. Liquid hydrogen has extremely low density (14 times lower than water) and requires extensive insulation—whilst gaseous hydrogen requires heavy tankage. Even when liquefied, hydrogen has a higher specific energy but the volumetric energetic storage is still roughly five times lower than gasoline. However, the energy density of hydrogen is considerably higher than that of electric batteries, making it a serious contender as an energy carrier to replace fossil fuels. The 'Hydrogen on Demand' process (see direct borohydride fuel cell) creates hydrogen as needed, but has other issues, such as the high price of the sodium borohydride that is the raw material.

Oxidizers

One-cylinder gasoline engine, c. 1910

Since air is plentiful at the surface of the earth, the oxidizer is typically atmospheric oxygen, which has the advantage of not being stored within the vehicle. This increases the power-to-weight and power-to-volume ratios. Other materials are used for special purposes, often to increase power output or to allow operation under water or in space.

- Compressed air has been commonly used in torpedoes.

- Compressed oxygen, as well as some compressed air, was used in the Japanese Type 93 torpedo. Some submarines carry pure oxygen. Rockets very often use liquid oxygen.

- Nitromethane is added to some racing and model fuels to increase power and control combustion.

- Nitrous oxide has been used—with extra gasoline—in tactical aircraft, and in specially equipped cars to allow short bursts of added power from engines that otherwise run on gasoline and air. It is also used in the Burt Rutan rocket spacecraft.

- Hydrogen peroxide power was under development for German World War II submarines. It may have been used in some non-nuclear submarines, and was used on some rocket engines (notably the Black Arrow and the Me-163 rocket plane).

- Other chemicals such as chlorine or fluorine have been used experimentally, but have not been found practical.

Cooling

Cooling is required to remove excessive heat — over heating can cause engine failure, usually from wear(due to heat-induced failure of lubrication), cracking or warping. Two most common forms of engine cooling are air-cooled and water-cooled. Most modern automotive engines are both water and air-cooled, as the water/liquid-coolant is carried to air-cooled fins and/or fans, whereas larger engines may be singularly water-cooled as they are stationary and have a constant supply of water through water-mains or fresh-water, while most power tool engines and other small engines are air-cooled. Some engines (air or water-cooled) also have an oil cooler. In some engines, especially for turbine engine blade cooling and liquid rocket engine cooling, fuel is used as a coolant, as it is simultaneously preheated before injecting it into a combustion chamber.

Starting

Electric Starter as used in automobiles

Internal Combustion engines must have their cycles started. In reciprocating engines this is accomplished by turning the crankshaft (Wankel Rotor Shaft) which induces the cycles of intake, compression, combustion, and exhaust. The first engines were started with a turn of their flywheels, while the first vehicle (the Daimler Reitwagen) was started with a hand crank. All ICE engined automobiles were started with hand cranks until Charles Kettering developed the electric starter for automobiles.

The most often found methods of starting ICE today is with an electric motor. As diesel engines have become larger another method has come into use as well, that is Air Starters.

Another method of starting is to use compressed air that is pumped into some cylinders of an engine to start it turning.

With two wheeled vehicles their engines may be started in three ways:

- By pedaling, as on a bicycle
- By pushing the vehicle and then engaging the clutch (Run and Bump Starting)
- Electric Starting

There are also starters where a spring is compressed by a crank motion and then used to start an engine. Small engines use a pull rope mechanism called recoil starting as the rope returns to storage after it has been pulled fully out to start the engine.

Turbine engines are frequently started by electric motor, or by air.

Measures of Engine Performance

Engine types vary greatly in a number of different ways:

- energy efficiency

- fuel/propellant consumption (brake specific fuel consumption for shaft engines, thrust specific fuel consumption for jet engines)

- power-to-weight ratio

- thrust to weight ratio

- Torque curves (for shaft engines) thrust lapse (jet engines)

- Compression ratio for piston engines, overall pressure ratio for jet engines and gas turbines

Energy Efficiency

Once ignited and burnt, the combustion products—hot gases—have more available thermal energy than the original compressed fuel-air mixture (which had higher chemical energy). The available energy is manifested as high temperature and pressure that can be translated into work by the engine. In a reciprocating engine, the high-pressure gases inside the cylinders drive the engine's pistons.

Once the available energy has been removed, the remaining hot gases are vented (often by opening a valve or exposing the exhaust outlet) and this allows the piston to return to its previous position (top dead center, or TDC). The piston can then proceed to the next phase of its cycle, which varies between engines. Any heat that is not translated into work is normally considered a waste product and is removed from the engine either by an air or liquid cooling system.

Internal combustion engines are heat engines, and as such their theoretical efficiency can be approximated by idealized thermodynamic cycles. The thermal efficiency of a theoretical cycle cannot exceed that of the Carnot cycle, whose efficiency is determined by the difference between the lower and upper operating temperatures of the engine. The upper operating temperature of an engine is limited by two main factors; the thermal operating limits of the materials, and the auto-ignition resistance of the fuel. All metals and alloys have a thermal operating limit, and there is significant research into ceramic materials that can be made with greater thermal stability and desirable structural properties. Higher thermal stability allows for a greater temperature difference between the lower (ambient) and upper operating temperatures, hence greater thermodynamic efficiency. Also, as the cylinder temperature rises, the engine becomes more prone to auto-ignition. This is caused when the cylinder temperature nears the flash point of the charge. At this point, ignition can spontaneously occur before the spark plug fires, causing excessive cylinder pressures. Auto-ignition can be mitigated by using fuels with high auto-ignition resistance (octane rating), however it still puts an upper bound on the allowable peak cylinder temperature.

The thermodynamic limits assume that the engine is operating under ideal conditions: a frictionless world, ideal gases, perfect insulators, and operation for infinite time. Real world applications introduce complexities that reduce efficiency. For example, a real engine runs best at a specific load, termed its power band. The engine in a car cruising on a highway is usually operating significantly below its ideal load, because it is designed for the higher loads required for rapid acceleration. In addition, factors such as wind resistance reduce overall system efficiency. Engine fuel economy is measured in miles per gallon or in liters per 100 kilometres. The volume of hydrocarbon assumes a standard energy content.

Most iron engines have a thermodynamic limit of 37%. Even when aided with turbochargers and stock efficiency aids, most engines retain an *average* efficiency of about 18%-20 %. The latest technologies in Formula One engines have seen a boost in thermal efficiency to almost 47%. Rocket engine efficiencies are much better, up to 70%, because they operate at very high temperatures and pressures and can have very high expansion ratios. Electric motors are better still, at around 85 -90 % efficiency or more, but they rely on an external power source (often another heat engine at a power plant subject to similar thermodynamic efficiency limits). However large stationary power plant turbines are typically significantly more efficient and cleaner than small mobile combustion engines in vehicles.

There are many inventions aimed at increasing the efficiency of IC engines. In general, practical engines are always compromised by trade-offs between different properties such as efficiency, weight, power, heat, response, exhaust emissions, or noise. Sometimes economy also plays a role in not only the cost of manufacturing the engine itself, but also manufacturing and distributing the fuel. Increasing the engine's efficiency brings better fuel economy but only if the fuel cost per energy content is the same.

Measures of Fuel Efficiency and Propellant Efficiency

For stationary and shaft engines including propeller engines, fuel consumption is measured by calculating the brake specific fuel consumption, which measures the mass flow rate of fuel consumption divided by the power produced.

For internal combustion engines in the form of jet engines, the power output varies drastically with airspeed and a less variable measure is used: thrust specific fuel consumption (TSFC), which is the mass of propellant needed to generate impulses that is measured in either pound force-hour or the grams of propellant needed to generate an impulse that measures one kilonewton-second.

For rockets, TSFC can be used, but typically other equivalent measures are traditionally used, such as specific impulse and effective exhaust velocity.

Air and Noise Pollution

Air Pollution

Internal combustion engines such as reciprocating internal combustion engines produce air pollution emissions, due to incomplete combustion of carbonaceous fuel. The main derivatives of the process are carbon dioxide CO_2, water and some soot — also called particulate matter (PM). The effects of inhaling particulate matter have been studied in humans and animals and include asthma, lung cancer, cardiovascular issues, and premature death. There are, however, some additional products of the combustion process that include nitrogen oxides and sulfur and some uncombusted hydrocarbons, depending on the operating conditions and the fuel-air ratio.

Not all of the fuel is completely consumed by the combustion process; a small amount of fuel is present after combustion, and some of it reacts to form oxygenates, such as formaldehyde or acetaldehyde, or hydrocarbons not originally present in the input fuel mixture. Incomplete combustion usually results from insufficient oxygen to achieve the perfect stoichiometric ratio. The flame is "quenched" by the relatively cool cylinder walls, leaving behind unreacted fuel that is expelled with the exhaust. When running at lower speeds, quenching is commonly observed in diesel (compression ignition) engines that run on natural gas. Quenching reduces efficiency and increases knocking, sometimes causing the engine to stall. Incomplete combustion also leads to the production of carbon monoxide (CO). Further chemicals released are benzene and 1,3-butadiene that are also hazardous air pollutants.

Increasing the amount of air in the engine reduces emissions of incomplete combustion products, but also promotes reaction between oxygen and nitrogen in the air to produce nitrogen oxides (NOx). NOx is hazardous to both plant and animal health, and leads to the production of ozone (O_3). Ozone is not emitted directly; rather, it is a secondary air pollutant, produced in the atmosphere by the reaction of NOx and volatile organic compounds in the presence of sunlight. Ground-level ozone is harmful to human health and the environment. Though the same chemical substance, ground-level ozone should not be confused with stratospheric ozone, or the ozone layer, which protects the earth from harmful ultraviolet rays.

Carbon fuels contain sulfur and impurities that eventually produce sulfur monoxides (SO) and sulfur dioxide (SO_2) in the exhaust, which promotes acid rain.

In the United States, nitrogen oxides, PM, carbon monoxide, sulphur dioxide, and ozone, are regulated as criteria air pollutants under the Clean Air Act to levels where human health and welfare are protected. Other pollutants, such as benzene and 1,3-butadiene, are regulated as hazardous air pollutants whose emissions must be lowered as much as possible depending on technological and practical considerations.

NOx, carbon monoxide and other pollutants are frequently controlled via exhaust gas recirculation which returns some of the exhaust back into the engine intake, and catalytic converters, which convert exhaust chemicals to harmless chemicals.

Non-road Engines

The emission standards used by many countries have special requirements for non-road engines which are used by equipment and vehicles that are not operated on the public roadways. The standards are separated from the road vehicles.

Noise Pollution

Significant contributions to noise pollution are made by internal combustion engines. Automobile and truck traffic operating on highways and street systems produce noise, as do aircraft flights due to jet noise, particularly supersonic-capable aircraft. Rocket engines create the most intense noise.

Idling

Internal combustion engines continue to consume fuel and emit pollutants when idling so it is desirable to keep periods of idling to a minimum. Many bus companies now instruct drivers to switch off the engine when the bus is waiting at a terminal.

In England, the Road Traffic Vehicle Emissions Fixed Penalty Regulations 2002 (Statutory Instrument 2002 No. 1808) introduced the concept of a *"stationary idling offence"*. This means that a driver can be ordered *"by an authorised person ... upon production of evidence of his authorisation, require him to stop the running of the engine of that vehicle"* and a *"person who fails to comply ... shall be guilty of an offence and be liable on summary conviction to a fine not exceeding level 3 on the standard scale"*. Only a few local authorities have implemented the regulations, one of them being Oxford City Council.

References

- Pulkrabek, Willard W. (1997). Engineering Fundamentals of the Internal Combustion Engine. Prentice Hall. p. 2. ISBN 9780135708545.

- "Turbulent times for Formula 1 engines result in unprecedented efficiency gains". Ars Technica. Retrieved 2016-05-20.

- "History of Technology: Internal Combustion engines". Encyclopædia Britannica. Britannica.com. Retrieved 2012-03-20.

- Takaishi, Tatsuo; Numata, Akira; Nakano, Ryouji; Sakaguchi, Katsuhiko (March 2008). "Approach to High Efficiency Diesel and Gas Engines" (PDF). Mitsubishi Heavy Industries Technical Review. 45 (1). Retrieved 2011-02-04.

- "CITY DEVELOPMENT - Fees & Charges 2010-11" (PDF). Oxford City Council. November 2011. Retrieved 2011-02-04.

- "New Benchmarks for Steam Turbine Efficiency - Power Engineering". Pepei.pennnet.com. 2010-08-24. Retrieved 2010-08-28.

- "The Road Traffic (Vehicle Emissions) (Fixed Penalty) (England) Regulations 2002". 195.99.1.70. 2010-07-16. Retrieved 2010-08-28.

2

Types of Internal Combustion Engine

Internal combustion engine is best understood in confluence with the major topics listed in the following chapter. The types of internal combustion engine explained in this chapter are diesel engine, petrol engine, four-stroke engine, two-stroke engine, jet engine, etc. The chapter strategically encompasses and incorporates the major components and key concepts of the internal combustion engine, providing a complete understanding.

Diesel Engine

The diesel engine (also known as a compression-ignition or CI engine) is an internal combustion engine in which ignition of the fuel that has been injected into the combustion chamber is caused by the high temperature which a gas achieves when greatly compressed (adiabatic compression). This contrasts with spark-ignition engines such as a petrol engine (gasoline engine) or gas engine (using a gaseous fuel as opposed to petrol), which use a spark plug to ignite an air-fuel mixture. In diesel engines, glow plugs (combustion chamber pre-warmers) may be used to aid starting in cold weather, or when the engine uses a lower compression-ratio, or both.

Diesel generator on an oil tanker

The diesel engine has the highest thermal efficiency (engine efficiency) of any practical internal or external combustion engine due to its very high expansion ratio and inherent lean burn which enables heat dissipation by the excess air. A small efficiency loss is also avoided compared to two-stroke non-direct-injection gasoline engines since unburnt fuel is not present at valve overlap and therefore no fuel goes directly from the intake/injection to the exhaust. Low-speed diesel engines (as used in ships and other applications where overall engine weight is relatively unimportant) can have a thermal efficiency that exceeds 50%.

A Diesel engine built by MAN AG in 1906

Diesel engines are manufactured in two-stroke and four-stroke versions. They were originally used as a more efficient replacement for stationary steam engines. Since the 1910s they have been used in submarines and ships. Use in locomotives, trucks, heavy equipment and electricity generation plants followed later. In the 1930s, they slowly began to be used in a few automobiles. Since the 1970s, the use of diesel engines in larger on-road and off-road vehicles in the US increased. According to the British Society of Motor Manufacturing and Traders, the EU average for diesel cars accounts for 50% of the total sold, including 70% in France and 38% in the UK.

The world's largest diesel engine is currently a Wärtsilä-Sulzer RTA96-C Common Rail marine diesel, which produces a peak power output of 84.42 MW (113,210 hp) at 102 rpm.

History

Hornsby-Akroyd oil engine

In 1885, the English inventor Herbert Akroyd Stuart began investigating the possibility of using paraffin oil (very similar to modern-day diesel) for an engine, which unlike petrol would be difficult to vaporise in a carburettor as its volatility is not sufficient to allow this.

His hot bulb engines, first prototyped in 1886 and built from 1891 by Richard Hornsby and Sons, used a pressurised fuel injection system. The Hornsby-Akroyd engine used a comparatively low compression ratio, so that the temperature of the air compressed in the combustion chamber at the end of the compression stroke was not high enough to initiate combustion. Combustion instead took place in a separated combustion chamber, the "vaporizer" or "hot bulb" mounted on the cylinder head, into which fuel was sprayed. Self-ignition occurred from contact between the fuel-air mixture and the hot walls of the vaporizer. As the engine's load increased, so did the temperature of the bulb, causing the ignition period to advance; to counteract pre-ignition, water was dripped into the air intake.

The modern Diesel engine incorporates the features of direct (airless) injection and compression-ignition. Both ideas were patented by Akroyd Stuart and Charles Richard Binney in May 1890. Another patent was taken out on 8 October 1890, detailing the working of a complete engine—essentially that of a diesel engine—where air and fuel are introduced separately. The difference between the Akroyd engine and the modern Diesel engine was the requirement to supply extra heat to the cylinder to start the engine from cold. By 1892, Akroyd Stuart had produced an updated version of the engine that no longer required the additional heat source, a year before Diesel's engine.

Diesel's original 1897 engine on display at the Deutsches Museum in Munich, Germany

In 1892, Akroyd Stuart patented a water-jacketed vaporiser to allow compression ratios to be increased. In the same year, Thomas Henry Barton at Hornsbys built a working high-compression version for experimental purposes, whereby the vaporiser was replaced with a cylinder head, therefore not relying on air being preheated, but by combustion through higher compression ratios. It ran for six hours—the first time automatic ignition was produced by compression alone. This was five years before Rudolf Diesel built his well-known high-compression prototype engine in 1897.

Rudolf Diesel was, however, subsequently credited with the compression ignition engine innovation, despite Akroyd-Stuart's engine being patented two years earlier. The higher compression and thermal efficiency is what distinguishes Diesel's patent, of 3,500 kilopascals (508 psi), from Ackroyd-Stuart's hot bulb compression ignition engine patent, of about 600 kilopascals (87 psi). Diesel improved his engine further, whereas Akroyd Stuart stopped development on his engine in 1893.

In 1892 Diesel received patents in Germany, Switzerland, the United Kingdom and the United States for "Method of and Apparatus for Converting Heat into Work". In 1893 he described a "slow-combustion engine" that first compressed air thereby raising its temperature above the igniting-point of the fuel, then gradually introducing fuel while letting the mixture expand "against resistance sufficiently to prevent an essential increase of temperature and pressure", then cutting off fuel and "expanding without transfer of heat". In 1894 and 1895 he filed patents and addenda in various countries for his Diesel engine; the first patents were issued in Spain (No. 16,654), France (No. 243,531) and Belgium (No. 113,139) in December 1894, and in Germany (No. 86,633) in 1895 and the United States (No. 608,845) in 1898. He operated his first successful engine in 1897.

At Augsburg, on August 10, 1893, Rudolf Diesel's prime model, a single 10-foot (3.0 m) iron cylinder with a flywheel at its base, ran on its own power for the first time. Diesel spent two more years making improvements and in 1896 demonstrated another model with a theoretical efficiency of 75%, in contrast to the 10% efficiency of the steam engine. By 1898, Diesel had become a millionaire. His engines were used to power pipelines, electric and water plants, automobiles and trucks, and marine craft. They were soon to be used in mines, oil fields, factories, and transoceanic shipping.

Timeline

- 1886: Herbert Akroyd Stuart builds a prototype hot bulb engine.

- 1891: Herbert Akroyd Stuart patents an internal combustion engine that uses a "hot bulb" and pressurized fuel injection.

- 1892: February 23, Rudolf Diesel obtained a patent (RP 67207) titled *"Arbeitsverfahren und Ausführungsart für Verbrennungsmaschinen"* (Working Methods and Techniques for Internal Combustion Engines).

- 1892: Akroyd Stuart builds his first working Diesel engine.

- 1893: Diesel's essay titled *Theory and Construction of a Rational Heat-engine to Replace the Steam Engine and Combustion Engines Known Today* appeared.

- 1893: August 10, Diesel built his first working prototype in Augsburg.

- 1896 Blackstone & Co, a Stamford farm implement they built lamp start oil engines.

- 1897: Adolphus Busch licenses rights to the Diesel Engine for the US and Canada.

- 1898: Diesel licensed his engine to Branobel, a Russian oil company interested in an engine that could consume non-distilled oil. Branobel's engineers spent four years designing a ship-mounted engine.

- 1899: Diesel licensed his engine to builders Krupp and Sulzer, who quickly became major manufacturers.

1900s

- 1902: Until 1910 MAN produced 82 copies of the stationary diesel engine.

- 1903: Two first diesel-powered ships were launched, both for river and canal operations: *La Petite-Pierre* in France, powered by Dyckhoff-built diesels, and *Vandal* tanker in Russia, powered by Swedish-built diesels with an electrical transmission.

- 1904: The French built the first diesel submarine, the Z.

- 1905: Four diesel engine turbochargers and intercoolers were manufactured by Büchl (CH), as well as a scroll-type supercharger from Creux (F) company.

- 1908: Prosper L'Orange and Deutz developed a precisely controlled injection pump with a needle injection nozzle.

- 1909: The prechamber with a hemispherical combustion chamber was developed by Prosper L'Orange with Benz.

1910s

- 1910: The Norwegian research ship *Fram* was a sailing ship fitted with an auxiliary diesel engine, and was thus the first ocean-going ship with a diesel engine.

- 1912: The Danish built the first ocean-going ship exclusively powered by a diesel engine, MS *Selandia*. The first locomotive with a diesel engine also appeared.

- 1913: US Navy submarines used NELSECO units. Rudolf Diesel died mysteriously when he crossed the English Channel on the SS *Dresden*.

- 1914: German U-boats were powered by MAN diesels.

- 1919: Prosper L'Orange obtained a patent on a prechamber insert and made a needle injection nozzle. First diesel engine from Cummins.

1920s

One of the eight-cylinder 3200 I.H.P. Harland and Wolff—Burmeister & Wain Diesel engines installed in the motorship *Glenapp*. This was the highest powered Diesel engine yet (1920) installed in a ship. Note man standing lower right for size comparison.

- 1921: Prosper L'Orange built a continuous variable output injection pump.

- 1922: The first vehicle with a (pre-chamber) diesel engine was Agricultural Tractor Type 6 of the Benz Söhne agricultural tractor OE Benz Sendling.

- 1923: The first truck with pre-chamber diesel engine made by MAN and Benz. Daimler-Motoren-Gesellschaft testing the first air-injection diesel-engined truck.

- 1924: The introduction on the truck market of the diesel engine by commercial truck manufacturers in the IAA. Fairbanks-Morse starts building diesel engines.

- 1927: First truck injection pump and injection nozzles of Bosch. First passenger car prototype of Stoewer.

1930s

- 1930s: Caterpillar started building diesels for their tractors.

- 1930: First US diesel-power passenger car (Cummins powered Packard) built in Columbus, Indiana (US).

- 1930: Beardmore Tornado diesel engines power the British airship R101.

- 1932: Introduction of the strongest diesel truck in the world by MAN with 160 hp (120 kW).

- 1933: First European passenger cars with diesel engines (Citroën Rosalie); Citroën used an engine of the English diesel pioneer Sir Harry Ricardo. The car did not go into production due to legal restrictions on the use of diesel engines.

- 1934: First turbo diesel engine for a railway train by Maybach. First streamlined, stainless steel passenger train in the US, the Pioneer Zephyr, using a Winton engine.

- 1934: First tank equipped with diesel engine, the Polish 7TP.

- 1934–35: Junkers Motorenwerke in Germany started production of the Jumo aviation diesel engine family, the most famous of these being the Jumo 205, of which over 900 examples were produced by the outbreak of World War II.

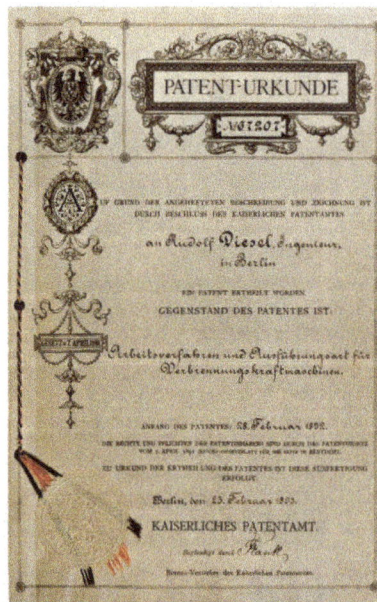

Rudolf Diesel's 1893 patent on his engine design

- 1936: Mercedes-Benz built the 260D diesel car. AT&SF inaugurated the diesel train Super Chief. The airship Hindenburg was powered by diesel engines. First series of passenger cars manufactured with diesel engine (Mercedes-Benz 260 D, Hanomag and Saurer). Daimler Benz airship diesel engine 602LOF6 for the LZ129 *Hindenburg* airship.

- 1937: The Soviet Union developed the Kharkiv model V-2 diesel engine, later used in the T-34 tanks, widely regarded as the best tank chassis of World War II.

- 1937: BMW 114 experimental airplane diesel engine development.

- 1938: General Motors forms the GM Diesel Division, later to become Detroit Diesel, and introduces the Series 71 inline high-speed medium-horsepower two stroke engine; GM's EMD subsidiary introduces the 567 two stroke medium-speed high-horsepower engine for locomotive, ship and stationary applications; These GM and EMD engines utilize GM's patented Unit injector.

- 1938: First turbo diesel engine of Saurer.

Fairbanks-Morse opposed piston diesel engines on the WWII submarine USS *Pampanito* (SS-383)
(on display in San Francisco)

1940s

- 1942: Tatra started production of Tatra 111 with air-cooled V12 diesel engine.

- 1943–46: The common-rail (CRD) system was invented (and patented by) Clessie Cummins

- 1944: Development of air cooling for diesel engines by Klöckner Humboldt Deutz AG (KHD) for the production stage, and later also for Magirus Deutz.

1950s

- 1953: Turbo-diesel truck for Mercedes in small series.

- 1954: Turbo-diesel truck in mass production by Volvo. First diesel engine with an overhead cam shaft of Daimler Benz.

- 1958 EMD introduces turbocharging for its 567 series of medium speed, high horsepower locomotive, stationary and marine engines. Every subsequent engine (645 and 710) would incorporate this turbocharger.

1960s

- 1960: The diesel drive displaced steam turbines and coal fired steam engines.

- 1962–65: A diesel compression braking system, eventually to be manufactured by Jacobs (of drill chuck fame) and nicknamed the "Jake Brake", was invented and patented by Clessie Cummins.

- 1968: Peugeot introduced the first 204 small cars with a transversally mounted diesel engine and front-wheel drive.

1970s

- 1973: DAF produced an air-cooled diesel engine.

- 1976 February: Tested a diesel engine for the Volkswagen Golf passenger car. The Cummins Common Rail injection system was further developed by the ETH Zurich from 1976 to 1992.

- 1978: Mercedes-Benz produced the first passenger car with a turbo-diesel engine (Mercedes-Benz 300 SD). Oldsmobile introduced the first passenger car diesel engine produced by an American car company.

- 1979: Peugeot 604, the first turbo-diesel car to be sold in Europe.

1980s

- 1985: ATI Intercooler diesel engine from DAF. European Truck Common Rail system with the IFA truck type W50 introduced.

- 1986: BMW 524td, the world's first passenger car equipped with an electronically controlled injection pump (developed by Bosch). The same year, the Fiat Croma was the first passenger car in the world to have a direct injection (turbocharged) diesel engine.

- 1987: Most powerful production truck with a 460 hp (340 kW) MAN diesel engine.

- 1989: Audi 100, the first passenger car in the world with a turbocharged direct injection and electronic control diesel engine.

1990s

- 1991: European emission standards Euro 1 met with the truck diesel engine of Scania.

- 1993: Pump nozzle injection introduced in Volvo truck engines.

- 1994: Unit injector system by Bosch for diesel engines. Mercedes-Benz unveils the first automotive diesel engine with four valves per cylinder. Medium speed high horsepower locomotive, ship and stationary diesel engines have utilized four valves per cylinder since at least 1938.

- 1995: First successful use of common rail in a production vehicle, by Denso in Japan, Hino "Rising Ranger" truck.

- 1996: First diesel engine with direct injection and four valves per cylinder, used in the Opel Vectra.

- 1997: First common rail diesel engine in a passenger car, the Alfa Romeo 156.

- 1998: BMW made history by winning the 24 Hour Nürburgring race with the 320d, powered by a two-litre, four-cylinder diesel engine. The combination of high-performance with better fuel efficiency allowed the team to make fewer pit stops during the long endurance race. Volkswagen introduces three and four-cylinder turbodiesel engines, with Bosch-developed electronically controlled unit injectors. Smart presented the first common rail three-cylinder diesel engine used in a passenger car (the Smart City Coupé).

- 1999: Euro 3 of Scania and the first common rail truck diesel engine of Renault.

2000s

- 2002: A street-driven Dodge Dakota pickup with a 735 horsepower (548 kW) diesel engine built at Gale banks engineering hauls its own service trailer to the Bonneville Salt Flats and set an FIA land speed record as the world's fastest pickup truck with a one-way run of 222 mph (357 km/h) and a two-way average of 217 mph (349 km/h).

- 2003: Piezoelectric injector technology by Bosch, Siemens and Delphi.

- 2004: In Western Europe, the proportion of passenger cars with diesel engine exceeded 50%. Selective catalytic reduction (SCR) system in Mercedes, Euro 4 with EGR system and particle filters of MAN. Audi A8 3.0 TDI is the first production vehicle in the world with common rail injection and piezoelectric injectors.

- 2006: Audi R10 TDI won the 12 Hours of Sebring and defeated all other engine concepts. The same car won the 2006 24 Hours of Le Mans. Euro 5 for all Iveco trucks. JCB Dieselmax broke the FIA diesel land speed record from 1973, eventually setting the new record at over 350 mph (563 km/h).

- 2007: Lombardini develops a new 440 cc twin-cyinder common rail diesel engine, which two years later sees application in automotive use, in the Ligier microcars. At the time, this engine was considered to be the smallest twin-cyinder engine with a common rail system.

- 2008: Subaru introduced the first horizontally opposed diesel engine to be fitted to a passenger car. This is a Euro 5 compliant engine with an EGR system. SEAT wins the drivers' title and the manufacturers' title in the FIA World Touring Car Championship with the SEAT León TDI. The achievements are repeated in the following season.

- 2009: Volkswagen won the 2009 Dakar Rally held in Argentina and Chile. The first diesel to do so. Race Touareg 2 models finished first and second. The same year, Volvo is claimed the world's strongest truck with their FH16 700. An inline 6-cylinder, 16 L (976 cu in) 700 hp (522 kW) diesel engine producing 3150 Nm (2323.32 lb•ft) of torque and fully complying with Euro 5 emission standards.

2010s

- 2010: Mitsubishi developed and started mass production of its 4N13 1.8 L DOHC I4, the world's first passenger car diesel engine that features a variable valve timing system. Scania

AB's V8 had the highest torque and power ratings of any truck engine, 730 hp (544 kW) and 3,500 N·m (2,581 ft·lb).

- 2011: Piaggio launches a twin-cyinder turbodiesel engine, with common rail injection, on its new range of microvans.

- 2012: Common rail systems working with pressures of 2,500 bar launched.

Operating Principle

p-V Diagram for the Ideal Diesel cycle. The cycle follows the numbers 1–4 in clockwise direction. The horizontal axis is Volume of the cylinder. In the diesel cycle the combustion occurs at almost constant pressure. On this diagram the work that is generated for each cycle corresponds to the area within the loop.

Diesel engine model, left side

Diesel engine model, right side

The diesel internal combustion engine differs from the gasoline powered Otto cycle by using highly compressed hot air to ignite the fuel rather than using a spark plug (*compression ignition* rather than *spark ignition*).

In the true diesel engine, only air is initially introduced into the combustion chamber. The air is then compressed with a compression ratio typically between 15:1 and 23:1. This high compression causes the temperature of the air to rise. At about the top of the compression stroke, fuel is injected directly into the compressed air in the combustion chamber. This may be into a (typically toroidal) void in the top of the piston or a *pre-chamber* depending upon the design of the engine. The fuel injector ensures that the fuel is broken down into small droplets, and that the fuel is distributed evenly. The heat of the compressed air vaporizes fuel from the surface of the droplets. The vapour is then ignited by the heat from the compressed air in the combustion chamber, the droplets continue to vaporise from their surfaces and burn, getting smaller, until all the fuel in the droplets has been burnt. Combustion occurs at a substantially constant pressure during the initial part of the power stroke. The start of vaporisation causes a delay before ignition and the characteristic diesel knocking sound as the vapour reaches ignition temperature and causes an abrupt increase in pressure above the piston (not shown on the P-V indicator diagram). When combustion is complete the combustion gases expand as the piston descends further; the high pressure in the cylinder drives the piston downward, supplying power to the crankshaft.

As well as the high level of compression allowing combustion to take place without a separate ignition system, a high compression ratio greatly increases the engine's efficiency. Increasing the compression ratio in a spark-ignition engine where fuel and air are mixed before entry to the cylinder is limited by the need to prevent damaging pre-ignition. Since only air is compressed in a diesel engine, and fuel is not introduced into the cylinder until shortly before top dead centre (TDC), premature detonation is not a problem and compression ratios are much higher.

The p–V diagam is a simplified and idealised representation of the events involved in a Diesel engine cycle, arranged to illustrate the similarity with a Carnot cycle. Starting at 1, the piston is at bottom dead centre and both valves are closed at the start of the compression stroke; the cylinder contains air at atmospheric pressure. Between 1 and 2 the air is compressed adiabatically—that is without heat transfer to or from the environment—by the rising piston. (This is only approximately true since there will be some heat exchange with the cylinder walls.) During this compression, the volume is reduced, the pressure and temperature both rise. At or slightly before 2 (TDC) fuel is injected and burns in the compressed hot air. Chemical energy is released and this constitutes an injection of thermal energy (heat) into the compressed gas. Combustion and heating occur between 2 and 3. In this interval the pressure remains constant since the piston descends, and the volume increases; the temperature rises as a consequence of the energy of combustion. At 3 fuel injection and combustion are complete, and the cylinder contains gas at a higher temperature than at 2. Between 3 and 4 this hot gas expands, again approximately adiabatically. Work is done on the system to which the engine is connected. During this expansion phase the volume of the gas rises, and its temperature and pressure both fall. At 4 the exhaust valve opens, and the pressure falls abruptly to atmospheric (approximately). This is unresisted expansion and no useful work is done by it. Ideally the adiabatic expansion should continue, extending the line 3–4 to the right until the pressure falls to that of the surrounding air, but the loss of efficiency caused by this unresisted expansion is justified by the practical difficulties involved in recovering it (the engine would have to be much larger). After the opening of the exhaust valve, the exhaust stroke follows, but this (and the following induction stroke) are not shown on the diagram. If shown, they would be represented by a low-pressure loop at the bottom of the diagram. At 1 it is assumed that the exhaust and induction strokes have been completed, and the cylinder is again filled with air. The piston-cylinder system absorbs energy between 1 and 2—this is the work needed to compress the air in the cylinder, and is provided by mechanical kinetic energy stored in the flywheel of the engine. Work output is done by the piston-cylinder combination between 2 and 4. The difference between these two increments of work is the indicated work output per cycle, and is represented by the area enclosed by the p–V loop. The adiabatic expansion is in a higher pressure range than that of the compression because the gas in the cylinder is hotter during expansion than during compression. It is for this reason that the loop has a finite area, and the net output of work during a cycle is positive.

Early Fuel Injection Systems

Diesel's original engine injected fuel with the assistance of compressed air, which atomized the fuel and forced it into the engine through a nozzle (a similar principle to an aerosol spray). The nozzle opening was closed by a pin valve lifted by the camshaft to initiate the fuel injection before top dead centre (TDC). This is called an air-blast injection. Driving the three stage compressor used some power but the efficiency and net power output was more than any other combustion engine at that time.

Diesel engines in service today raise the fuel to extreme pressures by mechanical pumps and deliver it to the combustion chamber by pressure-activated injectors without compressed air. With direct injected diesels, injectors spray fuel through 4 to 12 small orifices in its nozzle. The early air injection diesels always had a superior combustion without the sharp increase in pressure during combustion. Research is now being performed and patents are being taken out to again use some form of air injection to reduce the nitrogen oxides and pollution, reverting to Diesel's original implementation with its superior combustion and possibly quieter operation. In all major aspects, the modern diesel engine holds true to Rudolf Diesel's original design, that of igniting fuel by compression at an extremely high pressure within the cylinder. With much higher pressures and high technology injectors, present-day diesel engines use the so-called solid injection system applied by Herbert Akroyd Stuart for his hot bulb engine. The indirect injection engine could be considered the latest development of these low speed *hot bulb* ignition engines.

Fuel Delivery

Diesel engines are also produced with two significantly different injection locations. "Direct" and "Indirect". Indirect injected engines place the injector in a pre-combustion chamber in the head which due to thermal losses generally require a "glow plug" to start and very high compression ratio, usually in the range of 21:1 to 23:1 ratio. Direct injected engines use a generally donut shaped combustion chamber void on the top of the piston. Thermal efficiency losses are significantly lower in DI engines which facilitates a much lower compression ratio generally between 14:1 and 20:1 but most DI engines are closer to 17:1. The direct injected process is significantly more internally violent and thus requires careful design, and more robust construction. The lower compression ratio also creates challenges for emissions due to partial burn. Turbocharging is particularly suited to DI engines since the low compression ratio facilitates meaningful forced induction, and the increase in airflow allows capturing additional fuel efficiency not only from more complete combustion, but also from lowering parasitic efficiency losses when properly operated, by widening both power and efficiency curves. The violent combustion process of direct injection also creates more noise, but modern designs using "split shot" injectors or similar multi shot processes have dramatically amended this issue by firing a small charge of fuel before the main delivery which pre-charges the combustion chamber for a less abrupt and in most cases slightly cleaner burn.

A vital component of all diesel engines is a mechanical or electronic governor which regulates the idling speed and maximum speed of the engine by controlling the rate of fuel delivery. Unlike Otto-cycle engines, incoming air is not throttled and a diesel engine without a governor cannot have a stable idling speed and can easily overspeed, resulting in its destruction. Mechanically governed fuel injection systems are driven by the engine's gear train. These systems use a combination of springs and weights to control fuel delivery relative to both load and speed. Modern electronically controlled diesel engines control fuel delivery by use of an electronic control module (ECM) or electronic control unit (ECU). The ECM/ECU receives an engine speed signal, as well as other operating parameters such as intake manifold pressure and fuel temperature, from a sensor and controls the amount of fuel and start of injection timing through actuators to maximise power and efficiency and minimise emissions. Controlling the timing of the start of injection of fuel into the cylinder is a key to minimizing emissions, and maximizing fuel economy (efficiency), of the engine. The timing is measured in degrees of crank angle of the piston before top dead centre. For example, if the ECM/ECU initiates fuel injection when the piston is 10° before TDC, the start of injection, or timing, is said to be 10° BTDC. Optimal timing will depend on the engine design as well as its speed and load, and is usually 4° BTDC in 1,350–6,000 HP, net, "medium speed" locomotive, marine and stationary diesel engines.

Advancing the start of injection (injecting before the piston reaches to its SOI-TDC) results in higher in-cylinder pressure and temperature, and higher efficiency, but also results in increased engine noise due to faster cylinder pressure rise and increased oxides of nitrogen (NO_x) formation due to higher combustion temperatures. Delaying start of injection causes incomplete combustion, reduced fuel efficiency and an increase in exhaust smoke, containing a considerable amount of particulate matter and unburned hydrocarbons.

Major Advantages

Diesel engines have several advantages over other internal combustion engines:

- They burn less fuel than a petrol engine performing the same work, due to the engine's higher temperature of combustion and greater expansion ratio. Gasoline engines are typically 30% efficient while diesel engines can convert over 45% of the fuel energy into mechanical energy (see Carnot cycle for further explanation).

- They have no high voltage electrical ignition system, resulting in high reliability and easy adaptation to damp environments. The absence of coils, spark plug wires, etc., also eliminates a source of radio frequency emissions which can interfere with navigation and communication equipment, which is especially important in marine and aircraft applications, and for preventing interference with radio telescopes.

- The longevity of a diesel engine is generally about twice that of a petrol engine due to the increased strength of parts used. Diesel fuel has better lubrication properties than petrol as well. Indeed, in unit injectors, the fuel is employed for three distinct purposes: injector lubrication, injector cooling and injection for combustion.

- Diesel fuel is distilled directly from petroleum. Distillation yields some gasoline, but the yield would be inadequate without catalytic reforming, which is a more costly process.

- Although diesel fuel will burn in open air using a wick, it does not release a large amount of flammable vapor which could lead to an explosion. The low vapor pressure of diesel is especially advantageous in marine applications, where the accumulation of explosive fuel-air mixtures is a particular hazard. For the same reason, diesel engines are immune to vapor lock.

- For any given partial load the fuel efficiency (mass burned per energy produced) of a diesel engine remains nearly constant, as opposed to petrol and turbine engines which use proportionally more fuel with partial power outputs.

- They generate less waste heat in cooling and exhaust.

- Diesel engines can accept super- or turbo-charging pressure without any natural limit, constrained only by the strength of engine components. This is unlike petrol engines, which inevitably suffer detonation at higher pressure.

- The carbon monoxide content of the exhaust is minimal.

- Biodiesel is an easily synthesized, non-petroleum-based fuel (through transesterification) which can run directly in many diesel engines, while gasoline engines either need adaptation to run synthetic fuels or else use them as an additive to gasoline (e.g., ethanol added to gasohol).

Bus powered by biodiesel

Mechanical and Electronic Injection

Many configurations of fuel injection have been used over the course of the 20th century.

Most present-day diesel engines use a mechanical single plunger high-pressure fuel pump driven by the engine crankshaft. For each engine cylinder, the corresponding plunger in the fuel pump measures out the correct amount of fuel and determines the timing of each injection. These engines use injectors that are very precise spring-loaded valves that open and close at a specific fuel pressure. Separate high-pressure fuel lines connect the fuel pump with each cylinder. Fuel volume for each single combustion is controlled by a slanted groove in the plunger which rotates only a few degrees releasing the pressure and is controlled by a mechanical governor, consisting of weights rotating at engine speed constrained by springs and a lever. The injectors are held open by the fuel pressure. On high-speed engines the plunger pumps are together in one unit. The length of fuel lines from the pump to each injector is normally the same for each cylinder in order to obtain the same pressure delay.

A cheaper configuration on high-speed engines with fewer than six cylinders is to use an axial-piston distributor pump, consisting of one rotating pump plunger delivering fuel to a valve and line for each cylinder (functionally analogous to points and distributor cap on an Otto engine).

Many modern systems have a single fuel pump which supplies fuel constantly at high pressure with a common rail (single fuel line common) to each injector. Each injector has a solenoid operated by an electronic control unit, resulting in more accurate control of injector opening times that depend on other control conditions, such as engine speed and loading, and providing better engine performance and fuel economy.

Both mechanical and electronic injection systems can be used in either direct or indirect injection configurations.

Two-stroke diesel engines with mechanical injection pumps can be inadvertently run in reverse, albeit in a very inefficient manner, possibly damaging the engine. Large ship two-stroke diesels are designed to run in either direction, obviating the need for a gearbox.

Indirect Injection

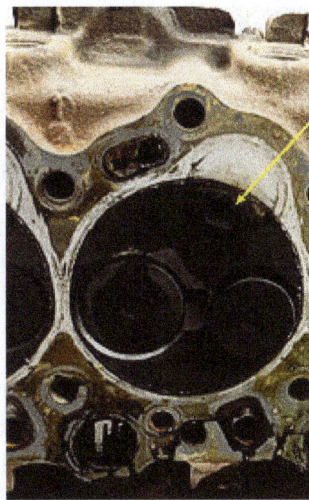

Arrow indicates opening from pre-chamber

An indirect injection diesel engine delivers fuel into a chamber off the combustion chamber, called a pre combustion chamber or ante-chamber, where combustion begins and then spreads into the main combustion chamber, assisted by turbulence created in the chamber. This system allows for a smoother, quieter running engine, and because combustion is assisted by turbulence, injector pressures can be lower, about 100 bar (10 MPa; 1,500 psi), using a single orifice tapered jet injector. Mechanical injection systems allowed high-speed running suitable for road vehicles (typically up to speeds of around 4,000 rpm). The pre-chamber had the disadvantage of increasing heat loss to the engine's cooling system, and restricting the combustion burn, which reduced the efficiency by 5–10%. Indirect injection engines are cheaper to build and it is easier to produce smooth, quiet-running vehicles with a simple mechanical system. In road-going vehicles most prefer the greater efficiency and better controlled emission levels of direct injection. Indirect injection diesels can still be found in the many ATV diesel applications.

Direct Injection

Direct injection diesel engines have injectors mounted at the top of the combustion chamber. The injectors are activated using one of two methods - hydraulic pressure from the fuel pump, or an electronic signal from an engine controller.

Hydraulic pressure activated injectors can produce harsh engine noise. Fuel consumption is about 15–20% lower than indirect injection diesels. The extra noise is generally not a problem for industrial uses of the engine, but for automotive usage, buyers have to decide whether or not the increased fuel efficiency would compensate for the extra noise. The noise level of modern common rail direct injection diesel engines with multiple injections per combustion cycle is however comparable to those of their gasoline counterparts.

Electronic control of the fuel injection transformed the direct injection engine by allowing much greater control over the combustion.

Unit Direct Injection

Unit direct injection also injects fuel directly into the cylinder of the engine. In this system the injector and the pump are combined into one unit positioned over each cylinder controlled by the camshaft. Each cylinder has its own unit eliminating the high-pressure fuel lines, achieving a more consistent injection. This type of injection system, also developed by Bosch, is used by Volkswagen AG in cars (where it is called a *Pumpe-Düse-System*—literally *pump-nozzle system*) and by Mercedes-Benz ("PLD") and most major diesel engine manufacturers in large commercial engines (MAN SE, CAT, Cummins, Detroit Diesel, Electro-Motive Diesel, Volvo). With recent advancements, the pump pressure has been raised to 2,400 bars (240 MPa; 35,000 psi), allowing injection parameters similar to common rail systems.

Common Rail Direct Injection

In common rail systems, the separate pulsing high-pressure fuel line to each cylinder's injector is also eliminated. Instead, a high-pressure pump pressurizes fuel at up to 2,500 bar (250 MPa; 36,000 psi), in a "common rail". The common rail is a tube that supplies each computer-controlled injector containing a precision-machined nozzle and a plunger driven by a solenoid or piezoelectric actuator.

Cold Weather

Starting

In cold weather, high speed diesel engines can be difficult to start because the mass of the cylinder block and cylinder head absorb the heat of compression, preventing ignition due to the higher surface-to-volume ratio. Pre-chambered engines make use of small electric heaters inside the pre-chambers called glowplugs, while direct-injected engines have these glowplugs in the combustion chamber.

Many engines use resistive heaters in the intake manifold to warm the inlet air for starting, or until the engine reaches operating temperature. Engine block heaters (electric resistive heaters in the engine block) connected to the utility grid are used in cold climates when an engine is turned off for extended periods (more than an hour), to reduce startup time and engine wear. Block heaters are also used for emergency power standby Diesel-powered generators which must rapidly pick up load on a power failure. In the past, a wider variety of cold-start methods were used. Some engines, such as Detroit Diesel engines used a system to introduce small amounts of ether into the inlet manifold to start combustion. Others used a mixed system, with a resistive heater burning methanol. An impromptu method, particularly on out-of-tune engines, is to manually spray an aerosol can of ether-based engine starter fluid into the intake air stream (usually through the intake air filter assembly).

Gelling

Diesel fuel is also prone to *waxing* or *gelling* in cold weather; both are terms for the solidification of diesel oil into a partially crystalline state. The crystals build up in the fuel line (especially in fuel filters), eventually starving the engine of fuel and causing it to stop running. Low-output electric heaters in fuel tanks and around fuel lines are used to solve this problem. Also, most engines have a *spill return* system, by which any excess fuel from the injector pump and injectors is returned to the fuel tank. Once the engine has warmed, returning warm fuel prevents waxing in the tank.

Due to improvements in fuel technology with additives, waxing rarely occurs in all but the coldest weather when a mix of diesel and kerosene may be used to run a vehicle. Gas stations in regions with a cold climate are required to offer winterized diesel in the cold seasons that allow operation below a specific Cold Filter Plugging Point. In Europe these diesel characteristics are described in the EN 590 standard.

Supercharging and Turbocharging

Most diesels are now turbocharged and some are both turbo charged and supercharged. Because diesels do not have fuel in the cylinder before combustion is initiated, more than one bar (100 kPa) of air can be loaded in the cylinder without preignition. A turbocharged engine can produce significantly more power than a naturally aspirated engine of the same configuration, as having more air in the cylinders allows more fuel to be burned and thus more power to be produced. A supercharger is powered mechanically by the engine's crankshaft, while a turbocharger is powered by the engine exhaust, not requiring any mechanical power. Turbocharging can improve the fuel economy of diesel engines by recovering waste heat from the exhaust, increasing the excess air factor, and increasing the ratio of engine output to friction losses.

A two-stroke engine does not have a discrete exhaust and intake stroke and thus is incapable of self-aspiration. Therefore, all two-stroke engines must be fitted with a blower to charge the cylinders with air and assist in dispersing exhaust gases, a process referred to as scavenging. In some cases, the engine may also be fitted with a turbocharger, whose output is directed into the blower inlet.

A few designs employ a hybrid turbocharger (a turbo-compressor system) for scavenging and charging the cylinders, which device is mechanically driven at cranking and low speeds to act as a blower, but which acts as a true turbocharger at higher speeds and loads. A hybrid turbocharger can revert to compressor mode during commands for large increases in engine output power.

As turbocharged or supercharged engines produce more power for a given engine size as compared to naturally aspirated engines, attention must be paid to the mechanical design of components, lubrication, and cooling to handle the power. Pistons are usually cooled with lubrication oil sprayed on the bottom of the piston. Large engines may use water, sea water, or oil supplied through telescoping pipes attached to the crosshead.

Types

Size Groups

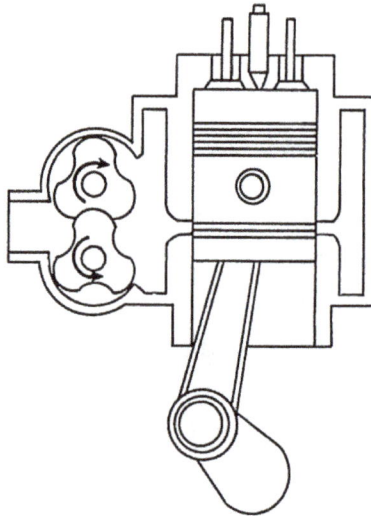

Two Cycle Diesel engine with Roots blower, typical of Detroit Diesel and some Electro-Motive Diesel Engines

There are three size groups of Diesel engines

- Small—under 188 kW (252 hp) output

- Medium

- Large

Basic Types

There are two basic types of Diesel Engines

- Four stroke cycle

- Two stroke cycle

Early Engines

Rudolf Diesel based his engine on Nikolaus Otto's 1876 engine, with the goal of improving its efficiency. Diesel's engine concepts were set forth in patents in 1892 and 1893. As such, diesel engines in the late 19th and early 20th centuries used the same basic layout and form as industrial steam engines, with long-bore cylinders, external valve gear, cross-head bearings and an open crankshaft connected to a large flywheel. Smaller engines would be built with vertical cylinders, while most medium- and large-sized industrial engines were built with horizontal cylinders, just as steam engines had been. Engines could be built with more than one cylinder in both cases. The largest early diesels resembled the triple-expansion steam reciprocating engine, being tens of feet high with vertical cylinders arranged in-line. These early engines ran at very slow speeds—partly due to the limitations of their air-blast injector equipment and partly so they would be compatible with the majority of industrial equipment designed for steam engines; maximum speeds of 100–300 rpm were common. Engines were usually started by allowing compressed air into the cylinders to turn the engine, although smaller engines could be started by hand.

In 1897, when the first Diesel engine was completed Adolphus Busch traveled to Cologne and negotiated exclusive right to produce the Diesel engine in the US and Canada. In his examination of the engine, it was noted that the Diesel at that time operated at thermodynamic efficiencies of 32–35%, while a typical triple expansion steam engine would operate at about 18%.

In the early decades of the 20th century, when large diesel engines were first being used, the engines took a form similar to the compound steam engines common at the time, with the piston being connected to the connecting rod by a crosshead bearing. Following steam engine practice some manufacturers made double-acting two-stroke and four-stroke diesel engines to increase power output, with combustion taking place on both sides of the piston, with two sets of valve gear and fuel injection. While it produced large amounts of power and was very efficient, the double-acting diesel engine's main problem was producing a good seal where the piston rod passed through the bottom of the lower combustion chamber to the crosshead bearing, and no more were built. By the 1930s turbochargers were fitted to some engines. Crosshead bearings are still used to reduce the wear on the cylinders in large long-stroke main marine engines.

Modern High and Medium-speed Engines

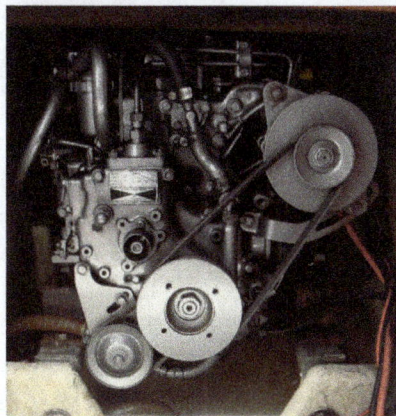

A Yanmar 2GM20 marine diesel engine, installed in a sailboat

As with petrol engines, there are two classes of diesel engines in current use: two-stroke and four-stroke. The four-stroke type is the "classic" version, tracing its lineage back to Rudolf Diesel's prototype. It is also the most commonly used form, being the preferred power source for many motor vehicles, especially buses and trucks. Much larger engines, such as used for railroad locomotion and marine propulsion, are often two-stroke units, offering a more favourable power-to-weight ratio, as well as better fuel economy. The most powerful engines in the world are two-stroke diesels of mammoth dimensions.

Two-stroke diesel engine operation is similar to that of petrol counterparts, except that fuel is not mixed with air before induction, and the crankcase does not take an active role in the cycle. The traditional two-stroke design relies upon a mechanically driven positive displacement blower to charge the cylinders with air before compression and ignition. The charging process also assists in expelling (scavenging) combustion gases remaining from the previous power stroke.

The archetype of the modern form of the two-stroke diesel is the (high-speed) Detroit Diesel Series 71 engine, designed by Charles F. "Boss" Kettering and his colleagues at General Motors Corporation in 1938, in which the blower pressurizes a chamber in the engine block that is often referred to as the "air box". The (very much larger medium-speed) Electro-Motive Diesel engine is used as the prime mover in EMD diesel-electric locomotive, marine and stationary applications, and was designed by the same team, and is built to the same principle. However, a significant improvement built into most later EMD engines is the mechanically assisted turbo-compressor, which provides charge air using mechanical assistance during starting (thereby obviating the necessity for Roots-blown scavenging), and provides charge air using an exhaust gas-driven turbine during normal operations—thereby providing true turbocharging and additionally increasing the engine's power output by at least fifty percent.

Three English Electric 7SRL diesel-alternator sets being installed at the Saateni Power Station, Zanzibar 1955

In a two-stroke diesel engine, as the cylinder's piston approaches the bottom dead centre exhaust ports or valves are opened relieving most of the excess pressure after which a passage between the air box and the cylinder is opened, permitting air flow into the cylinder. The air flow blows the remaining combustion gases from the cylinder—this is the scavenging process. As the piston passes through bottom centre and starts upward, the passage is closed and compression commences, culminating in fuel injection and ignition. Refer to two-stroke diesel engines for more detailed coverage of aspiration types and supercharging of two-stroke diesel engines.

Normally, the number of cylinders are used in multiples of two, although any number of cylinders can be used as long as the load on the crankshaft is counterbalanced to prevent excessive vibration. The inline-six-cylinder design is the most prolific in light- to medium-duty engines, though small V8 and larger inline-four displacement engines are also common. Small-capacity engines (generally considered to be those below five litres in capacity) are generally four- or six-cylinder types, with the four-cylinder being the most common type found in automotive uses. Five-cylinder diesel engines have also been produced, being a compromise between the smooth running of the six-cylinder and the space-efficient dimensions of the four-cylinder. Diesel engines for smaller plant machinery, boats, tractors, generators and pumps may be four, three or two-cylinder types, with the single-cylinder diesel engine remaining for light stationary work. Direct reversible two-stroke marine diesels need at least three cylinders for reliable restarting forwards and reverse, while four-stroke diesels need at least six cylinders.

The desire to improve the diesel engine's power-to-weight ratio produced several novel cylinder arrangements to extract more power from a given capacity. The uniflow opposed-piston engine uses two pistons in one cylinder with the combustion cavity in the middle and gas in- and outlets at the ends. This makes a comparatively light, powerful, swiftly running and economic engine suitable for use in aviation. An example is the Junkers Jumo 204/205. The Napier Deltic engine, with three cylinders arranged in a triangular formation, each containing two opposed pistons, the whole engine having three crankshafts, is one of the better known.

Modern Low-speed Engines

Low-speed diesel engines (as used in ships and other applications where overall engine weight is relatively unimportant) often have a thermal efficiency which exceeds 50%.

Gas Generator

Before 1950, Sulzer started experimenting with two-stroke engines with boost pressures as high as 6 atmospheres, in which all the output power was taken from an exhaust gas turbine. The two-stroke pistons directly drove air compressor pistons to make a positive displacement gas generator. Opposed pistons were connected by linkages instead of crankshafts. Several of these units could be connected to provide power gas to one large output turbine. The overall thermal efficiency was roughly twice that of a simple gas turbine. This system was derived from Raúl Pateras Pescara's work on free-piston engines in the 1930s.

Advantages and Disadvantages Versus Spark-ignition Engines

Fuel Economy

The MAN S80ME-C7 low speed diesel engines use 155 grams (5.5 oz) of fuel per kWh for an overall energy conversion efficiency of 54.4%, which is the highest conversion of fuel into power by any single-cycle internal or external combustion engine (The efficiency of a combined cycle gas turbine system can exceed 60%.) Diesel engines are more efficient than gasoline (petrol) engines of the same power rating, resulting in lower fuel consumption. A common margin is 40% more miles per gallon for an efficient turbodiesel. For example, the current model Škoda Octavia, using Volkswagen Group engines, has a combined Euro rating of 6.2 L/100 km (46 mpg$_{-imp}$; 38 mpg$_{US}$) for the 102 bhp (76 kW) petrol engine and 4.4 L/100 km (64 mpg$_{-imp}$; 53 mpg$_{US}$) for the 105 bhp (78 kW) diesel engine.

However, such a comparison does not take into account that diesel fuel is denser and contains about 15% more energy by volume. Although the calorific value of the fuel is slightly lower at 45.3 MJ/kg (megajoules per kilogram) than petrol at 45.8 MJ/kg, liquid diesel fuel is significantly denser than liquid petrol. This is significant because volume of fuel, in addition to mass, is an important consideration in mobile applications.

Adjusting the numbers to account for the energy density of diesel fuel, the overall energy efficiency is still about 20% greater for the diesel version.

While a higher compression ratio is helpful in raising efficiency, diesel engines are much more efficient than gasoline (petrol) engines when at low power and at engine idle. Unlike the petrol engine, diesels lack a butterfly valve (throttle) in the inlet system, which closes at idle. This creates parasitic loss and destruction of availability of the incoming air, reducing the efficiency of petrol engines at idle. In many applications, such as marine, agriculture, and railways, diesels are left idling and unattended for many hours, sometimes even days. These advantages are especially attractive in locomotives (see dieselisation).

Even though diesel engines have a theoretical fuel efficiency of 75%, in practice it is lower. Engines in large diesel trucks, buses, and newer diesel cars can achieve peak efficiencies around 45%, and could reach 55% efficiency in the near future. However, average efficiency over a driving cycle is lower than peak efficiency. For example, it might be 37% for an engine with a peak efficiency of 44%.

Torque

Diesel engines produce more torque than petrol engines for a given displacement due to their higher compression ratio. Higher pressure in the cylinder and higher forces on the connecting rods and crankshaft require stronger, heavier components. Heavier rotating components prevent diesel engines from revving as high as petrol engines for a given displacement. Diesel engines generally have similar power and inferior power to weight ratios as compared to petrol engines. Petrol engines must be geared lower to get the same torque as a comparable diesel but since petrol engines rev higher both will have similar acceleration. An arbitrary amount of torque at the wheels can be gained by gearing any power source down sufficiently (including a hand crank). For example, a theoretical engine with a constant 200 ft/lbs of torque and a 3000 rpm rev limit has just as much power (a little over 114 hp) as another theoretical engine with a constant maximum 100 ft/lbs of torque and a 6000 rpm rev limit. A (lossless) 2 to 1 reduction gear on the second engine will output a constant maximum 200 ft/lbs of torque at a maximum of 3000 rpm, with no change in power. Comparing engines based on (maximum) torque is just as useful as comparing them based on (maximum) rpm.

Power

Conditions in the diesel engine differ from the spark-ignition engine due to the different thermodynamic cycle. In addition the power and engine speed are directly controlled by the fuel supply, rather than by controlling the air supply as in an otto cycle engine.

The average diesel engine has a poorer power-to-weight ratio than the petrol engine. This is because the diesel must operate at lower engine speeds due to the need for heavier, stronger parts to resist the operating pressure caused by the high compression ratio of the engine, which increases the forces on the parts due to inertial forces. In addition, diesels are often built with stronger parts to give them longer lives and better reliability, important considerations in industrial applications.

Diesel engines usually have longer stroke lengths chiefly to facilitate achieving the necessary compression ratios. As a result, piston and connecting rods are heavier and more force must be transmitted through the connecting rods and crankshaft to change the momentum of the piston. This is another reason that a diesel engine must be stronger for the same power output as a petrol engine.

Yet it is this characteristic that has allowed some enthusiasts to acquire significant power increases with turbocharged engines by making fairly simple and inexpensive modifications. A petrol engine of similar size cannot put out a comparable power increase without extensive alterations because the stock components cannot withstand the higher stresses placed upon them. Since a diesel engine is already built to withstand higher levels of stress, it makes an ideal candidate for performance tuning at little expense. However, it should be said that any modification that raises the amount of fuel and air put through a diesel engine will increase its operating temperature, which will reduce its life and increase service requirements. These are issues with newer, lighter, *high-performance* diesel engines which are not "overbuilt" to the degree of older engines and they are being pushed to provide greater power in smaller engines.

Forced Induction

The addition of a turbocharger will increase fuel economy and power, while a supercharger will increase power only by mitigating the fuel-air intake speed limit mentioned above for a given engine displacement. Boost pressures can be higher on diesels than on petrol engines, due to the latter's susceptibility to knock, and the higher compression ratio allows a diesel engine to be more efficient than a comparable spark ignition engine. Because the burned gases are expanded further in a diesel engine cylinder, the exhaust gas is cooler, meaning turbochargers require less cooling, and can be more reliable, than in spark-ignition engines.

Without the risk of knocking, boost pressure in a diesel engine can be much higher; it is possible to run as much boost as the engine will physically stand before breaking apart.

A combination of improved mechanical technology (such as multi-stage injectors which fire a short "pilot charge" of fuel into the cylinder to warm the combustion chamber before delivering the main fuel charge), higher injection pressures that have improved the atomization of fuel into smaller droplets, and electronic control (which can adjust the timing and length of the injection process to optimize it for all speeds and temperatures) have mitigated most of these problems in the latest generation of common-rail designs, while greatly improving engine efficiency. Poor power and narrow torque bands have been addressed by superchargers, turbochargers, (especially variable geometry turbochargers), intercoolers, and a large efficiency increase from about 35% to 45% for the latest engines in the last 15 years.

Emissions

Since the diesel engine uses less fuel than the petrol engine per unit distance, the diesel produces less carbon dioxide (CO_2) per unit distance. Recent advances in production and changes in the political climate have increased the availability and awareness of biodiesel, an alternative to petroleum-derived diesel fuel with a much lower net-sum emission of CO_2, due to the absorption of CO_2 by plants used to produce the fuel. However, the use of waste vegetable oil, sawmill waste from managed forests in Finland, and advances in the production of vegetable oil from algae demonstrate great promise in providing feed stocks for sustainable biodiesel that are not in competition with food production.

When a diesel engine runs at low power, there is enough oxygen present to burn the fuel—diesel engines only make significant amounts of carbon monoxide when running under a load.

Diesel fuel is injected just before the power stroke. As a result, the fuel cannot burn completely unless it has a sufficient amount of oxygen. This can result in incomplete combustion and black smoke in the exhaust if more fuel is injected than there is air available for the combustion process. Modern engines with electronic fuel delivery can adjust the timing and amount of fuel delivered, and so operate with less waste of fuel. In a mechanical system fuel timing system, the injection and duration must be set to be efficient at the anticipated operating rpm and load, and so the settings are less than ideal when the engine is running at any other RPM. The electronic injection can "sense" engine revs, load, even boost and temperature, and continuously alter the timing to match the given situation. In the petrol engine, air and fuel are mixed for the entire compression stroke, ensuring complete mixing even at higher engine speeds.

Diesel exhaust is well known for its characteristic smell, but this smell in recent years has become much less due to use of low sulfur fuel.

Diesel exhaust has been found to contain a long list of toxic air contaminants. Among these pollutants, fine particle pollution is an important as a cause of diesel's harmful health effects. However, when diesel engines burn their fuel with high oxygen levels, this results in high combustion temperatures and higher efficiency, and these particles tend to burn, but the amount of NOx pollution tends to increase.

NOx pollution can be reduced with diesel exhaust fluid, which is injected into the exhaust stream, and catalytically destroys the NOx chemical species. Exhaust gas recirculation which works by recirculating a portion of an engine's exhaust gas back to the engine cylinders also has very positive effects on NOx emissions.

Noise

The distinctive noise of a diesel engine is variably called diesel clatter, diesel nailing, or diesel knock. Diesel clatter is caused largely by the diesel combustion process; the sudden ignition of the diesel fuel when injected into the combustion chamber causes a pressure wave. Engine designers can reduce diesel clatter through: indirect injection; pilot or pre-injection; injection timing; injection rate; compression ratio; turbo boost; and exhaust gas recirculation (EGR). Common rail diesel injection systems permit multiple injection events as an aid to noise reduction. Diesel fuels with a higher cetane rating modify the combustion process and reduce diesel clatter. CN (Cetane number) can be raised by distilling higher quality crude oil, by catalyzing a higher quality product or by using a cetane improving additive.

A combination of improved mechanical technology such as multi-stage injectors which fire a short "pilot charge" of fuel into the cylinder to initiate combustion before delivering the main fuel charge, higher injection pressures that have improved the atomisation of fuel into smaller droplets, and electronic control (which can adjust the timing and length of the injection process to optimise it for all speeds and temperatures), have partially mitigated these problems in the latest generation of common-rail designs, while improving engine efficiency.

Reliability

For most industrial or nautical applications, reliability is considered more important than light weight and high power.

The lack of an electrical ignition system greatly improves the reliability. The high durability of a diesel engine is also due to its overbuilt nature (see above). Diesel fuel is a better lubricant than petrol and thus, it is less harmful to the oil film on piston rings and cylinder bores as occurs in petro powered engines; it is routine for diesel engines to cover 400,000 km (250,000 mi) or more without a rebuild.

Due to the greater compression ratio and the increased weight of the stronger components, starting a diesel engine is harder than starting a gasoline engine of similar design and displacement. More torque from the starter motor is required to push the engine through the compression cycle when starting compared to a petrol engine. This can cause difficulty when starting in winter time if using conventional automotive batteries because of the lower current available.

Either an electrical starter or an air-start system is used to start the engine turning. On large engines, pre-lubrication and slow turning of an engine, as well as heating, are required to minimise the amount of engine damage during initial start-up and running. Some smaller military diesels can be started with an explosive cartridge, called a Coffman starter, which provides the extra power required to get the machine turning. In the past, Caterpillar and John Deere used a small petrol *pony* engine in their tractors to start the primary diesel engine. The pony engine heated the diesel to aid in ignition and used a small clutch and transmission to spin up the diesel engine. Even more unusual was an International Harvester design in which the diesel engine had its own carburetor and ignition system, and started on petrol. Once warmed, the operator moved two levers to switch the engine to diesel operation, and work could begin. These engines had very complex cylinder heads, with their own petrol combustion chambers, and were vulnerable to expensive damage if special care was not taken (especially in letting the engine cool before turning it off).

Cylinder Cavitation and Erosion Damage

One phenomenon that can affect water-cooled diesel engines is cylinder cavitation and erosion. This is due to a phenomenon in high-compression engines where the ignition of the fuel in the cylinder causes a high-frequency vibration that causes bubbles to form in the coolant in contact with the cylinder. When these tiny bubbles collapse, coolant impacts the cylinder wall, over time causing small holes to form in the cylinder wall. This damage is mitigated in some engines with coatings, or with a coolant additive specifically designed to prevent cavitation and erosion damage. Engines damaged in this way will require the affected cylinder to be repaired (where possible) or will be rendered unusable.

Quality and Variety of Fuels

Petrol/gasoline engines are limited in the variety and quality of the fuels they can burn. Older petrol engines fitted with a carburetor required a volatile fuel that would vaporise easily to create the necessary air-fuel ratio for combustion. Because both air and fuel are admitted to the cylinder, if the compression ratio of the engine is too high or the fuel too volatile (with too low an octane rating), the fuel will ignite under compression, as in a diesel engine, before the piston reaches the top of its stroke. This pre-ignition causes a power loss and over time major damage to the piston and cylinder. The need for a fuel that is volatile enough to vaporise but not too volatile (to avoid pre-ignition) means that petrol engines will only run on a narrow range of fuels. There has been some success at dual-fuel engines that use petrol and ethanol, petrol and propane, and petrol and methane.

In diesel engines, a mechanical injector system vaporizes the fuel directly into the combustion chamber or a pre-combustion chamber (as opposed to a Venturi jet in a carburetor, or a fuel injector in a fuel injection system vaporising fuel into the intake manifold or intake runners as in a petrol engine). This *forced vaporisation* means that less-volatile fuels can be used. More crucially, because only air is inducted into the cylinder in a diesel engine, the compression ratio can be much higher as there is no risk of pre-ignition provided the injection process is accurately timed. This means that cylinder temperatures are much higher in a diesel engine than a petrol engine, allowing less volatile fuels to be used.

Diesel fuel is a form of light fuel oil, very similar to kerosene (paraffin), but diesel engines, especially older or simple designs that lack precision electronic injection systems, can run on a wide variety of other fuels. Some of the most common alternatives are Jet A-1 type jet fuel or vegetable oil from a very wide variety of plants. Some engines can be run on vegetable oil without modification, and most others require fairly basic alterations. Biodiesel is a pure diesel-like fuel refined from vegetable oil and can be used in nearly all diesel engines. Requirements for fuels to be used in diesel engines are the ability of the fuel to flow along the fuel lines, the ability of the fuel to lubricate the injector pump and injectors adequately, and its ignition qualities (ignition delay, cetane number). Inline mechanical injector pumps generally tolerate poor-quality or bio-fuels better than distributor-type pumps. Also, indirect injection engines generally run more satisfactorily on bio-fuels than direct injection engines. This is partly because an indirect injection engine has a much greater 'swirl' effect, improving vaporisation and combustion of fuel, and because (in the case of vegetable oil-type fuels) lipid depositions can condense on the cylinder walls of a direct-injection engine if combustion temperatures are too low (such as starting the engine from cold).

It is often reported that Diesel designed his engine to run on peanut oil, but this is false. Patent number 608845 describes his engine as being designed to run on pulverulent solid fuel (coal dust). Diesel stated in his published papers, "at the Paris Exhibition in 1900 (*Exposition Universelle*) there was shown by the Otto Company a small diesel engine, which, at the request of the French Government ran on Arachide (earth-nut or peanut) oil (see biodiesel), and worked so smoothly that only a few people were aware of it. The engine was constructed for using mineral oil, and was then worked on vegetable oil without any alterations being made. The French Government at the time thought of testing the applicability to power production of the Arachide, or earth-nut, which grows in considerable quantities in their African colonies, and can easily be cultivated there." Diesel himself later conducted related tests and appeared supportive of the idea.

Most large marine diesels run on heavy fuel oil (sometimes called "bunker oil"), which is a thick, viscous and almost flameproof fuel which is very safe to store and cheap to buy in bulk as it is a waste product from the petroleum refining industry. The fuel must not only be pre-heated, but must be kept heated during handling and storage in order to maintain its pumpability. This is usually accomplished by steam tracing on fuel lines and steam coils in fuel oil tanks. The fuel is then preheated to over 100C before entering the engine in order to attain the proper viscosity for atomisation.

Fuel and Fluid Characteristics

Diesel engines can operate on a variety of different fuels, depending on configuration, though the eponymous diesel fuel derived from crude oil is most common. The engines can work with the full spectrum of crude oil distillates, from natural gas, alcohols, petrol, wood gas to the *fuel oils* from diesel oil to residual fuels. Many automotive diesel engines would run on 100% biodiesel without any modifications.

The type of fuel used is selected to meet a combination of service requirements, and fuel costs. Good-quality diesel fuel can be synthesised from vegetable oil and alcohol. Diesel fuel can be made from coal or other carbon base using the Fischer–Tropsch process. Biodiesel is growing in popularity since it can frequently be used in unmodified engines, though production remains limited. Recently, biodiesel from coconut, which can produce a very promising coco methyl ester (CME), has characteristics which enhance lubricity and combustion giving a regular diesel engine without any modification more power, less particulate matter or black smoke, and smoother engine performance. The Philippines pioneers in the research on Coconut based CME with the help of German and American scientists. Petroleum-derived diesel is often called *petrodiesel* if there is need to distinguish the source of the fuel.

Pure plant oils are increasingly being used as a fuel for cars, trucks and remote combined heat and power generation especially in Germany where hundreds of decentralised small- and medium-sized oil presses cold press oilseed, mainly rapeseed, for fuel. There is a Deutsches Institut für Normung fuel standard for rapeseed oil fuel.

Residual fuels are the "dregs" of the distillation process and are a thicker, heavier oil, or oil with higher viscosity, which are so thick that they are not readily pumpable unless heated. Residual fuel oils are cheaper than clean, refined diesel oil, although they are dirtier. Their main considerations are for use in ships and very large generation sets, due to the cost of the large volume of fuel consumed, frequently amounting to many tonnes per hour. The poorly refined biofuels straight vegetable oil (SVO) and waste vegetable oil (WVO) can fall into this category, but can be viable fuels on non-common rail or TDI PD diesels with the simple conversion of fuel heating to 80 to 100 degrees Celsius to reduce viscosity, and adequate filtration to OEM standards. Engines using these heavy oils have to start and shut down on standard diesel fuel, as these fuels will not flow through fuel lines at low temperatures. Moving beyond that, use of low-grade fuels can lead to serious maintenance problems because of their high sulphur and lower lubrication properties. Most diesel engines that power ships like supertankers are built so that the engine can safely use low-grade fuels due to their separate cylinder and crankcase lubrication.

Normal diesel fuel is more difficult to ignite and slower in developing fire than petrol because of its higher flash point, but once burning, a diesel fire can be fierce.

Fuel contaminants such as dirt and water are often more problematic in diesel engines than in petrol engines. Water can cause serious damage, due to corrosion, to the injection pump and injectors; and dirt, even very fine particulate matter, can damage the injection pumps due to the close tolerances that the pumps are machined to. All diesel engines will have a fuel filter (usually much finer than a filter on a petrol engine), and a water trap. The water trap (which is sometimes part of the fuel filter) often has a float connected to a warning light, which warns when there is too much water in the trap, and must be drained before damage to the engine can result. The fuel filter must be replaced much more often on a diesel engine than on a petrol engine, changing the fuel filter every 2–4 oil changes is not uncommon for some vehicles.

Safety

Fuel Flammability

Diesel fuel is less flammable than petrol, leading to a lower risk of fire caused by fuel in a vehicle equipped with a diesel engine.

In yachts, diesel engines are often used because the petrol (gasoline) that fuels spark-ignition engines releases combustible vapors which can lead to an explosion if it accumulates in a confined space such as the bottom of a vessel. Ventilation systems are mandatory on petrol-powered vessels.

The United States Army and NATO use only diesel engines and turbines because of fire hazard. Although neither gasoline nor diesel is explosive in liquid form, both can create an explosive air/vapor mix under the right conditions. However, diesel fuel is less prone due to its lower vapor pressure, which is an indication of evaporation rate. The Material Safety Data Sheet for ultra-low sulfur diesel fuel indicates a vapor explosion hazard for diesel indoors, outdoors, or in sewers.

US Army gasoline-engined tanks during World War II were nicknamed Ronsons, because of their greater likelihood of catching fire when damaged by enemy fire, although tank fires were usually caused by detonation of the ammunition rather than fuel, while diesel tanks such as the Soviet T-34 were less prone to catching fire.

Maintenance Hazards

Fuel injection introduces potential hazards in engine maintenance due to the high fuel pressures used. Residual pressure can remain in the fuel lines long after an injection-equipped engine has been shut down. This residual pressure must be relieved, and if it is done so by external bleed-off, the fuel must be safely contained. If a high-pressure diesel fuel injector is removed from its seat and operated in open air, there is a risk to the operator of injury by hypodermic jet-injection, even with only 100 pounds per square inch (690 kPa) pressure. The first known such injury occurred in 1937 during a diesel engine maintenance operation.

Cancer

Diesel exhaust has been classified as an IARC Group 1 carcinogen. It causes lung cancer and is associated with an increased risk for bladder cancer.

Applications

The characteristics of diesel have different advantages for different applications.

Passenger Cars

Diesel engines have long been popular in bigger cars and have been used in smaller cars such as superminis like the Peugeot 205, in Europe since the 1980s. Diesel engines tend to be more economical at regular driving speeds and are much better at city speeds. Their reliability and life-span tend to be better (as detailed). Some 40% or more of all cars sold in Europe are diesel-powered where they are considered a low CO_2 option. Mercedes-Benz in conjunction with Robert Bosch GmbH produced diesel-powered passenger cars starting in 1936 and very large numbers are used all over the world (often as "Grande Taxis" in the Third World). Diesel-powered passenger cars are very popular in India too, since the price of diesel fuel there is lower as compared to petrol. As a result, predominantly petrol-powered car manufacturers including the Japanese car manufacturers produce and market diesel-powered cars in India. Diesel-powered cars also dominate the Indian taxi industry.

Railroad Rolling Stock

Diesel engines have eclipsed steam engines as the prime mover on all non-electrified railroads in the industrialized world. The first diesel locomotives appeared in the early 20th century, and diesel multiple units soon after. While electric locomotives have replaced the diesel locomotive for some passenger traffic in Europe and Asia, diesel is still today very popular for cargo-hauling freight trains and on tracks where electrification is not feasible. Most modern diesel locomotives are actually diesel-electric locomotives: the diesel engine is used to power an electric generator that in turn powers electric traction motors with no mechanical connection between diesel engine and traction. After 2000, environmental requirements has caused higher development cost for engines, and it has become common for passenger multiple units to use engines and automatic mechanical gearboxes made for trucks. Up to four such combinations might be used to get enough power in a train.

Other Transport Uses

Larger transport applications (trucks, buses, etc.) also benefit from the Diesel's reliability and high torque output. Diesel displaced paraffin (or tractor vaporising oil, TVO) in most parts of the world by the end of the 1950s with the US following some 20 years later.

- Aircraft

- Marine

- Motorcycles

In merchant ships and boats, the same advantages apply with the relative safety of Diesel fuel an additional benefit. The German pocket battleships were the largest Diesel warships, but the German torpedo-boats known as E-boats (*Schnellboot*) of the Second World War were also Diesel craft. Conventional submarines have used them since before World War I, relying on the almost total absence of carbon monoxide in the exhaust. American World War II Diesel-electric submarines operated on two-stroke cycle, as opposed to the four-stroke cycle that other navies used.

Non-road Diesel Engines

Non-road diesel engines include mobile equipment and vehicles that are not used on the public roadways such as construction equipment and agricultural tractors.

Military Fuel Standardisation

NATO has a single vehicle fuel policy and has selected diesel for this purpose. The use of a single fuel simplifies wartime logistics. NATO and the United States Marine Corps have even been developing a diesel military motorcycle based on a Kawasaki off road motorcycle the KLR 650, with a purpose designed naturally aspirated direct injection diesel at Cranfield University in England, to be produced in the US, because motorcycles were the last remaining gasoline-powered vehicle in their inventory. Before this, a few civilian motorcycles had been built using adapted stationary diesel engines, but the weight and cost disadvantages generally outweighed the efficiency gains.

Non-transport Uses

A 1944 V12 2,300 kW power plant undergoing testing & restoration

Diesel engines are also used to power permanent, portable, and backup generators, irrigation pumps, corn grinders, and coffee de-pulpers.

Engine Speeds

Within the diesel engine industry, engines are often categorized by their rotational speeds into three unofficial groups:

- High-speed engines (> 1,000 rpm),

- Medium-speed engines (300–1,000 rpm), and

- Slow-speed engines (< 300 rpm).

High- and medium-speed engines are predominantly four-stroke engines; except for the Detroit Diesel two-stroke range. Medium-speed engines are physically larger than high-speed engines and can burn lower-grade (slower-burning) fuel than high-speed engines. Slow-speed engines are predominantly large two-stroke crosshead engines, hence very different from high- and medi-um-speed engines. Due to the lower rotational speed of slow- and medium-speed engines, there is more time for combustion during the power stroke of the cycle, allowing the use of slower-burning fuels than high-speed engines.

High-speed Engines

High-speed (approximately 1,000 rpm and greater) engines are used to power trucks (lorries), buses, tractors, cars, yachts, compressors, pumps and small electrical generators. As of 2008, most high-speed engines have direct injection. Many modern engines, particularly in on-highway appli-cations, have common rail direct injection, which is cleaner burning.

Medium-speed Engines

Medium-speed engines are used in large electrical generators, ship propulsion and mechanical drive applications such as large compressors or pumps. Medium speed diesel engines operate on either diesel fuel or heavy fuel oil by direct injection in the same manner as low-speed engines.

Engines used in electrical generators run at approximately 300 to 1000 rpm and are optimized to run at a set synchronous speed depending on the generation frequency (50 or 60 hertz) and provide a rapid response to load changes. Typical synchronous speeds for modern medium-speed engines are 500/514 rpm (50/60 Hz), 600 rpm (both 50 and 60 Hz), 720/750 rpm, and 900/1000 rpm.

As of 2009, the largest medium-speed engines in current production have outputs up to approx-imately 20 MW (27,000 hp) and are supplied by companies like MAN B&W, Wärtsilä, and Rolls-Royce (who acquired Ulstein Bergen Diesel in 1999). Most medium-speed engines produced are four-stroke machines, however there are some two-stroke medium-speed engines such as by EMD (Electro-Motive Diesel), and the Fairbanks Morse OP (Opposed-piston engine) type.

Typical cylinder bore size for medium-speed engines ranges from 20 cm to 50 cm, and engine con-figurations typically are offered ranging from in-line 4-cylinder units to V-configuration 20-cylin-der units. Most larger medium-speed engines are started with compressed air direct on pistons, using an air distributor, as opposed to a pneumatic starting motor acting on the flywheel, which tends to be used for smaller engines. There is no definitive engine size cut-off point for this.

It should also be noted that most major manufacturers of medium-speed engines make natural gas-fueled versions of their diesel engines, which in fact operate on the Otto cycle, and require spark ignition, typically provided with a spark plug. There are also dual (diesel/natural gas/coal gas) fuel versions of medium and low speed diesel engines using a lean fuel air mixture and a small injection of diesel fuel (so-called "pilot fuel") for ignition. In case of a gas supply failure or maxi-mum power demand these engines will instantly switch back to full diesel fuel operation.

Low-speed Engines

Also known as *slow-speed*, or traditionally *oil engines*, the largest diesel engines are primar-ily used to power ships, although there are a few land-based power generation units as well. These extremely large two-stroke engines have power outputs up to approximately 85 MW (114,000 hp), operate in the range from approximately 60 to 200 rpm and are up to 15 m

(50 ft) tall, and can weigh over 2,000 short tons (1,800 t). They typically use direct injection running on cheap low-grade heavy fuel, also known as bunker C fuel, which requires heating in the ship for tanking and before injection due to the fuel's high viscosity. Often, the waste heat recovery steam boilers attached to the engine exhaust ducting generate the heat required for fuel heating. Provided the heavy fuel system is kept warm and circulating, engines can be started and stopped on heavy fuel.

The MAN B&W 5S50MC 5-cylinder, 2-stroke, low-speed marine diesel engine. This particular engine is found aboard a 29,000 tonne chemical carrier.

Large and medium marine engines are started with compressed air directly applied to the pistons. Air is applied to cylinders to start the engine forwards or backwards because they are normally directly connected to the propeller without clutch or gearbox, and to provide reverse propulsion either the engine must be run backwards or the ship will use an adjustable propeller. At least three cylinders are required with two-stroke engines and at least six cylinders with four-stroke engines to provide torque every 120 degrees.

Companies such as MAN B&W Diesel, and Wärtsilä design such large low-speed engines. They are unusually narrow and tall due to the addition of a crosshead bearing. As of 2007, the 14-cylinder Wärtsilä-Sulzer 14RTFLEX96-C turbocharged two-stroke diesel engine built by Wärtsilä licensee Doosan in Korea is the most powerful diesel engine put into service, with a cylinder bore of 960 mm (37.8 in) delivering 114,800 hp (85.6 MW). It was put into service in September 2006, aboard what was then the world's largest container ship *Emma Maersk* which belongs to the A.P. Moller-Maersk Group. Typical bore size for low-speed engines ranges from approximately 35 to 98 cm (14 to 39 in). As of 2008, all produced low-speed engines with crosshead bearings are in-line configurations; no Vee versions have been produced.

Current and Future Developments

As of 2008, many common rail and unit injection systems already employ new injectors using stacked piezoelectric wafers in lieu of a solenoid, giving finer control of the injection event.

Variable geometry turbochargers have flexible vanes, which move and let more air into the engine depending on load. This technology increases both performance and fuel economy. Boost lag is reduced as turbo impeller inertia is compensated for.

Accelerometer pilot control (APC) uses an accelerometer to provide feedback on the engine's level of noise and vibration and thus instruct the ECU to inject the minimum amount of fuel that will produce quiet combustion and still provide the required power (especially while idling).

The next generation of common rail diesels is expected to use variable injection geometry, which allows the amount of fuel injected to be varied over a wider range, and variable valve timing (see Mitsubishi's 4N13 diesel engine) similar to that of petrol engines. Particularly in the United States, coming tougher emissions regulations present a considerable challenge to diesel engine manufacturers. Ford's HyTrans Project has developed a system which starts the ignition in 400 ms, saving a significant amount of fuel on city routes, and there are other methods to achieve even more efficient combustion, such as homogeneous charge compression ignition, being studied.

Japanese and Swedish vehicle manufacturers are also developing diesel engines that run on dimethyl ether (DME).

Some recent diesel engine models utilize a copper alloy heat exchanger technology (CuproBraze) to take advantage of benefits in terms of thermal performance, heat transfer efficiency, strength/durability, corrosion resistance, and reduced emissions from higher operating temperatures.

Low Heat Rejection Engines

A special class of experimental prototype internal combustion piston engines has been developed over several decades with the goal of improving efficiency by reducing heat loss. These engines are variously called adiabatic engines; due to better approximation of adiabatic expansion; low heat rejection engines, or high temperature engines. They are generally piston engines with combustion chamber parts lined with ceramic thermal barrier coatings. Some make use of pistons and other parts made of titanium which has a low thermal conductivity and density. Some designs are able to eliminate the use of a cooling system and associated parasitic losses altogether. Developing lubricants able to withstand the higher temperatures involved has been a major barrier to commercialization.

Petrol Engine

A petrol engine (known as a gasoline engine in American English) is an internal combustion engine with spark-ignition, designed to run on petrol (gasoline) and similar volatile fuels. The first practical petrol engine was built in 1876 in Germany by Nikolaus August Otto, although there had been earlier attempts by Étienne Lenoir, Siegfried Marcus, Julius Hock and George Brayton. The first petrol combustion engine (one cylinder, 121.6 cm^3 displacement) was prototyped in 1882 in Italy by Enrico Bernardi. In most petrol engines, the fuel and air are usually pre-mixed before compression (although some modern petrol engines now use cylinder-direct petrol injection). The pre-mixing was formerly done in a carburetor, but now it is done by electronically controlled fuel injection, except in small engines where the cost/complication of electronics does not justify the added engine efficiency. The process differs from a diesel engine in the method of mixing the fuel and air, and in using spark plugs to initiate the combustion process. In a diesel engine, only air is compressed (and therefore heated), and the fuel is injected into very hot air at the end of the compression stroke, and self-ignites.

W16 petrol engine of the Bugatti Veyron

Compression Ratio

With both air and fuel in a closed cylinder, compressing the mixture too much poses the danger of auto-ignition — or behaving like a diesel engine. Because of the difference in burn rates between the two different fuels, petrol engines are mechanically designed with different timing than diesels, so to auto-ignite a petrol engine causes the expansion of gas inside the cylinder to reach its greatest point before the cylinder has reached the "top dead center" (TDC) position. Spark plugs are typically set statically or at idle at a minimum of 10 degrees or so of crankshaft rotation before the piston reaches TDC, but at much higher values at higher engine speeds to allow time for the fuel-air charge to substantially complete combustion before too much expansion has occurred - gas expansion occurring with the piston moving down in the power stroke. Higher octane petrol burns slower, therefore it has a lower propensity to auto-ignite and its rate of expansion is lower. Thus, engines designed to run high-octane fuel exclusively can achieve higher compression ratios.

Speed and Efficiency

Petrol engines run at higher speeds than diesels, partially due to their lighter pistons, connecting rods and crankshaft (a design efficiency made possible by lower compression ratios) and due to petrol burning more quickly than diesel. Because pistons in petrol engines tend to have much shorter strokes than pistons in diesel engines, typically it takes less time for a piston in a petrol engine to complete its stroke than a piston in a diesel engine. However the lower compression ratios of petrol engines give petrol engines lower efficiency than diesel engines.

Applications

Current

Petrol engines have many applications, including:

- Automobiles
- Motorcycles

- Aircraft

- Motorboats

- Small engines, such as lawn mowers, chainsaws and portable engine-generators

Historical

Before the use of diesel engines became widespread, petrol engines were used in buses, lorries (trucks) and a few railway locomotives. Examples:

- Bedford OB bus

- Bedford M series lorry

- GE 57-ton gas-electric boxcab locomotive

Design

Working Cycles

Petrol engines may run on the four-stroke cycle or the two-stroke cycle. For details of working cycles see:

- Four-stroke cycle

- Two-stroke cycle

- Wankel engine

Cylinder Arrangement

Common cylinder arrangements are from 1 to 6 cylinders in-line or from 2 to 16 cylinders in V-formation. Flat engines – like a V design flattened out – are common in small airplanes and motorcycles and were a hallmark of Volkswagen automobiles into the 1990s. Flat 6s are still used in many modern Porsches, as well as Subarus. Many flat engines are air-cooled. Less common, but notable in vehicles designed for high speeds is the W formation, similar to having 2 V engines side by side. Alternatives include rotary and radial engines the latter typically have 7 or 9 cylinders in a single ring, or 10 or 14 cylinders in two rings.

Cooling

Petrol engines may be air-cooled, with fins (to increase the surface area on the cylinders and cylinder head); or liquid-cooled, by a water jacket and radiator. The coolant was formerly water, but is now usually a mixture of water and either ethylene glycol or propylene glycol. These mixtures have lower freezing points and higher boiling points than pure water and also prevent corrosion, with modern antifreezes also containing lubricants and other additives to protect water pump seals and bearings. The cooling system is usually slightly pressurized to further raise the boiling point of the coolant.

Ignition

Petrol engines use spark ignition and high voltage current for the spark may be provided by a magneto or an ignition coil. In modern car engines the ignition timing is managed by an electronic Engine Control Unit.

Power Measurement

The most common way of engine rating is what is known as the brake power, measured at the flywheel, and given in kilowatts (metric) or horsepower (Imperial/USA). This is the actual mechanical power output of the engine in a usable and complete form. The term "brake" comes from the use of a brake in a dynamometer test to load the engine. For accuracy, it is important to understand what is meant by usable and complete. For example, for a car engine, apart from friction and thermodynamic losses inside the engine, power is absorbed by the water pump, alternator, and radiator fan, thus reducing the power available at the flywheel to move the car along. Power is also absorbed by the power steering pump and air conditioner (if fitted), but these are not installed for a power output test or calculation. Power output varies slightly according to the energy value of the fuel, the ambient air temperature and humidity, and the altitude. Therefore, there are agreed standards in the USA and Europe on the fuel to use when testing, and engines are rated at 25 °C (Europe), and 64 °F (USA) at sea level, 50% humidity. Marine engines, as supplied, usually have no radiator fan, and often no alternator. In such cases the quoted power rating does not allow for losses in the radiator fan and alternator. The SAE in USA, and the ISO in Europe publish standards on exact procedures, and how to apply corrections for deviating conditions like high altitude.

Car testers are most familiar with the chassis dynamometer or "rolling road" installed in many workshops. This measures drive wheel brake horsepower, which is generally 15-20% less than the brake horsepower measured at the crankshaft or flywheel on an engine dynamometer. A YouTube video shows workshop measurement of a car's power. The measured power curve in kW is shown at 3:39.

Four-stroke Engine

Four-stroke cycle used in gasoline/petrol engines. 1 = Intake, 2 = Compression, 3 = Power, 4 = Exhaust. The right blue side is the intake port and the left brown side is the exhaust port. The cylinder wall is a thin sleeve surrounding the piston head which creates a space for the combustion of fuel and the genesis of mechanical energy.

A four-stroke engine (also known as four cycle) is an internal combustion (IC) engine in which the piston completes four separate strokes while turning a crankshaft. A stroke refers to the full travel of the piston along the cylinder, in either direction. The four separate strokes are termed:

1. Intake: This stroke of the piston begins at top dead center (T.D.C.) and ends at bottom dead center (B.D.C.). In this stroke the intake valve must be in the open position while the piston pulls an air-fuel mixture into the cylinder by producing vacuum pressure into the cylinder through its downward motion.

2. Compression: This stroke begins at B.D.C, or just at the end of the suction stroke, and ends at T.D.C. In this stroke the piston compresses the air-fuel mixture in preparation for ignition during the power stroke (below). Both the intake and exhaust valves are closed during this stage.

3. Combustion: This is the start of the second revolution of the four stroke cycle. At this point the crankshaft has completed a full 360 degree revolution. While the piston is at T.D.C. (the end of the compression stroke) the compressed air-fuel mixture is ignited by a spark plug (in a gasoline engine) or by heat generated by high compression (diesel engines), forcefully returning the piston to B.D.C. This stroke produces mechanical work from the engine to turn the crankshaft.

4. Exhaust: During the *exhaust* stroke, the piston once again returns from B.D.C. to T.D.C. while the exhaust valve is open. This action expels the spent air-fuel mixture through the exhaust valve.

History

Otto Cycle

An Otto Engine from 1920's US Manufacture

Nikolaus August Otto as a young man was a traveling salesman for a grocery concern. In his travels he encountered the internal combustion engine built in Paris by Belgian expatriate Jean Joseph Etienne Lenoir. In 1860, Lenoir successfully created a double-acting engine that ran on illuminating gas at 4% efficiency. The 18 litre Lenoir Engine produced only 2 horsepower. The Lenoir engine ran on illuminating gas made from coal, which had been developed in Paris by Philip Lebon.

In testing a replica of the Lenoir engine in 1861 Otto became aware of the effects of compression on the fuel charge. In 1862, Otto attempted to produce an engine to improve on the poor efficiency and reliability of the Lenoir engine. He tried to create an engine that would compress the fuel mixture prior to ignition, but failed as that engine would run no more than a few minutes prior to its destruction. Many other engineers were trying to solve the problem, with no success.

In 1864, Otto and Eugen Langen founded the first internal combustion engine production company, NA Otto and Cie (NA Otto and Company). Otto and Cie succeeded in creating a successful atmospheric engine that same year. The factory ran out of space and was moved to the town of Deutz, Germany in 1869 where the company was renamed to Deutz

Gasmotorenfabrik AG (The Deutz Gas Engine Manufacturing Company). In 1872, Gottlieb Daimler was technical director and Wilhelm Maybach was the head of engine design. Daimler was a gunsmith who had worked on the Lenoir engine. By 1876, Otto and Langen succeeded in creating the first internal combustion engine that compressed the fuel mixture prior to combustion for far higher efficiency than any engine created to this time.

Daimler and Maybach left their employ at Otto and Cie and developed the first high-speed Otto engine in 1883. In 1885, they produced the first automobile to be equipped with an Otto engine. The Daimler *Reitwagen* used a hot-tube ignition system and the fuel known as Ligroin to become the world's first vehicle powered by an internal combustion engine. It used a four-stroke engine based on Otto's design. The following year Karl Benz produced a four-stroke engined automobile that is regarded as the first car.

In 1884, Otto's company, then known as Gasmotorenfabrik Deutz (GFD), developed electric ignition and the carburetor. In 1890, Daimler and Maybach formed a company known as Daimler Motoren Gesellschaft. Today, that company is Daimler-Benz.

Atkinson Cycle

This 2004 Toyota Prius hybrid has an Atkinson cycle engine as the petrol-electric hybrid engine

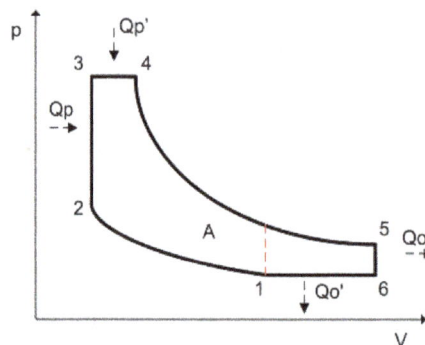

The Atkinson Gas Cycle

The Atkinson cycle engine is a type of single stroke internal combustion engine invented by James Atkinson in 1882. The Atkinson cycle is designed to provide efficiency at the expense of power density, and is used in some modern hybrid electric applications.

The original Atkinson cycle piston engine allowed the intake, compression, power, and exhaust strokes of the four-stroke cycle to occur in a single turn of the crankshaft and was designed to avoid infringing certain patents covering Otto cycle engines.

Due to the unique crankshaft design of the Atkinson, its expansion ratio can differ from its compression ratio and, with a power stroke longer than its compression stroke, the engine can achieve greater thermal efficiency than a traditional piston engine. While Atkinson's original design is no more than a historical curiosity, many modern engines use unconventional valve timing to produce the effect of a shorter compression stroke/longer power stroke, thus realizing the fuel economy improvements the Atkinson cycle can provide.

Diesel Cycle

Audi Diesel R15 at Le Mans

The diesel engine is a technical refinement of the 1876 Otto Cycle engine. Where Otto had realized in 1861 that the efficiency of the engine could be increased by first compressing the fuel mixture prior to its ignition, Rudolph Diesel wanted to develop a more efficient type of engine that could run on much heavier fuel. The Lenoir, Otto Atmospheric, and Otto Compression engines (both 1861 and 1876) were designed to run on Illuminating Gas (coal gas). With the same motivation as Otto, Diesel wanted to create an engine that would give small industrial concerns their own power source to enable them to compete against larger companies, and like Otto to get away from the requirement to be tied to a municipal fuel supply. Like Otto, it took more than a decade to produce the high compression engine that could self-ignite fuel sprayed into the cylinder. Diesel used an air spray combined with fuel in his first engine.

During initial development, one of the engines burst nearly killing him. He persisted and finally created an engine in 1893. The high compression engine, which ignites its fuel by the heat of compression is now called the Diesel engine whether a four-stroke or two-stroke design.

The four-stroke diesel engine has been used in the majority of heavy duty applications for many decades. It uses a heavy fuel containing more energy and requiring less refinement to produce. The most efficient Otto Cycle engines run near 30% efficiency.

Thermodynamic Analysis

The thermodynamic analysis of the actual four-stroke or two-stroke cycles is not a simple task. However, the analysis can be simplified significantly if air standard assumptions are utilized. The resulting cycle, which closely resembles the actual operating conditions, is the Otto cycle.

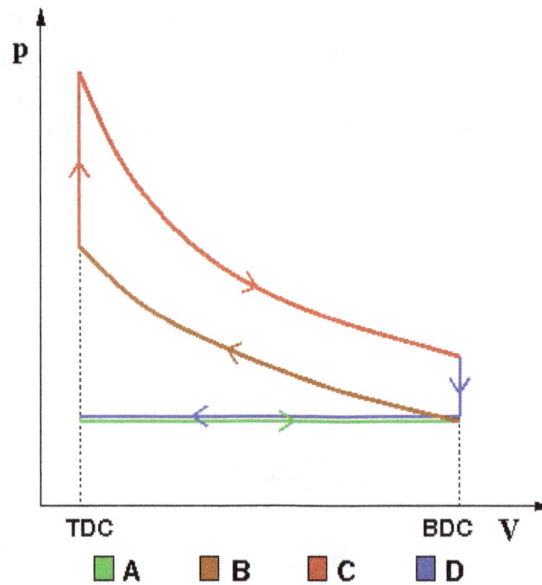

The idealized four-stroke Otto cycle p-V diagram: the intake (A) stroke is performed by an isobaric expansion, followed by the compression (B) stroke, performed by an adiabatic compression. Through the combustion of fuel an isochoric process is produced, followed by an adiabatic expansion, characterizing the power (C) stroke. The cycle is closed by an isochoric process and an isobaric compression, characterizing the exhaust (D) stroke.

During the normal operation of the engine as the fuel mixture is being compressed an electric arc is created to ignite the fuel. At low rpm this occurs close to TDC (Top Dead Centre). As engine rpm rises the spark point is moved earlier in the cycle so that the fuel charge can be ignited while it is still being compressed. We can see this advantage reflected in the various Otto engines designs. The atmospheric (non-compression) engine operated at 12% efficiency. The compressed charge engine had an operating efficiency of 30%.

Fuel Considerations

The problem with compressed charge engines is that the temperature rise of the compressed charge can cause pre-ignition. If this occurs at the wrong time and is too energetic, it can damage the engine. Different fractions of petroleum have widely varying flash points (the temperatures at which the fuel may self-ignite). This must be taken into account in engine and fuel design.

The tendency for the compressed fuel mixture to ignite early is limited by the chemical composition of the fuel. There are several grades of fuel to accommodate differing performance levels of engines. The fuel is altered to change its self ignition temperature. There are several ways to do this. As engines are designed with higher compression ratios the result is that pre-ignition is much more likely to occur since the fuel mixture is compressed to a higher temperature prior to deliberate ignition. The higher temperature more effectively evaporates fuels such as gasoline, which increases the efficiency of the compression engine. Higher Compression ratios also mean that the distance that the piston can push to produce power is greater (which is called the Expansion ratio).

The octane rating of a given fuel is a measure of the fuel's resistance to self-ignition. A fuel with a higher numerical octane rating allows for a higher compression ratio, which extracts more energy from the fuel and more effectively converts that energy into useful work while at the same time preventing engine damage from pre-ignition. High Octane fuel is also more expensive.

Diesel engines by their nature do not have concerns with pre-ignition. They have a concern with whether or not combustion can be started. The description of how likely Diesel fuel is to ignite is called the Cetane rating. Because Diesel fuels are of low volatility, they can be very hard to start when cold. Various techniques are used to start a cold Diesel engine, the most common being the use of a glow plug.

Design and Engineering Principles

Power Output Limitations

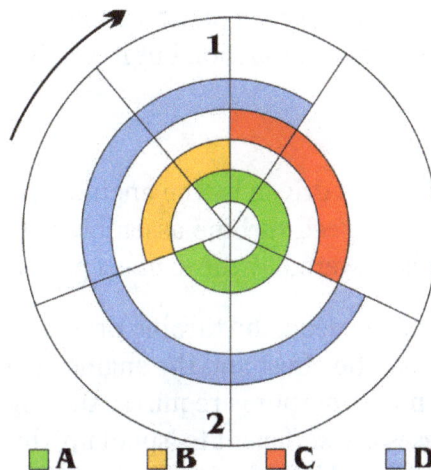

The four-stroke cycle1=TDC2=BDC A: Intake B: Compression C: Power D: Exhaust

The maximum amount of power generated by an engine is determined by the maximum amount of air ingested. The amount of power generated by a piston engine is related to its size (cylinder volume), whether it is a two-stroke or four-stroke design, volumetric efficiency, losses, air-to-fuel ratio, the calorific value of the fuel, oxygen content of the air and speed (RPM). The speed is ultimately limited by material strength and lubrication. Valves, pistons and connecting rods suffer severe acceleration forces. At high engine speed, physical breakage and piston ring flutter can occur, resulting in power loss or even engine destruction. Piston ring flutter occurs when the rings oscillate vertically within the piston grooves they reside in. Ring flutter compromises the seal between the ring and the cylinder wall, which causes a loss of cylinder pressure and power. If an engine spins too quickly, valve springs cannot act quickly enough to close the valves. This is commonly referred to as 'valve float', and it can result in piston to valve contact, severely damaging the engine. At high speeds the lubrication of piston cylinder wall interface tends to break down. This limits the piston speed for industrial engines to about 10 m/s.

Intake/Exhaust Port Flow

The output power of an engine is dependent on the ability of intake (air–fuel mixture) and exhaust matter to move quickly through valve ports, typically located in the cylinder head. To increase an engine's output power, irregularities in the intake and exhaust paths, such as casting flaws, can be removed, and, with the aid of an air flow bench, the radii of valve port turns and valve seat configuration can be modified to reduce resistance. This process is called porting, and it can be done by hand or with a CNC machine.

Supercharging

One way to increase engine power is to force more air into the cylinder so that more power can be produced from each power stroke. This can be done using some type of air compression device known as a supercharger, which can be powered by the engine crankshaft.

Supercharging increases the power output limits of an internal combustion engine relative to its displacement. Most commonly, the supercharger is always running, but there have been designs that allow it to be cut out or run at varying speeds (relative to engine speed). Mechanically driven supercharging has the disadvantage that some of the output power is used to drive the supercharger, while power is wasted in the high pressure exhaust, as the air has been compressed twice and then gains more potential volume in the combustion but it is only expanded in one stage.

Turbocharging

A turbocharger is a supercharger that is driven by the engine's exhaust gases, by means of a turbine. It consists of a two piece, high-speed turbine assembly with one side that compresses the intake air, and the other side that is powered by the exhaust gas outflow.

When idling, and at low-to-moderate speeds, the turbine produces little power from the small exhaust volume, the turbocharger has little effect and the engine operates nearly in a naturally aspirated manner. When much more power output is required, the engine speed and throttle opening are increased until the exhaust gases are sufficient to 'spool up' the turbocharger's turbine to start compressing much more air than normal into the intake manifold.

Turbocharging allows for more efficient engine operation because it is driven by exhaust pressure that would otherwise be (mostly) wasted, but there is a design limitation known as turbo lag. The increased engine power is not immediately available due to the need to sharply increase engine RPM, to build up pressure and to spin up the turbo, before the turbo starts to do any useful air compression. The increased intake volume causes increased exhaust and spins the turbo faster, and so forth until steady high power operation is reached. Another difficulty is that the higher exhaust pressure causes the exhaust gas to transfer more of its heat to the mechanical parts of the engine.

Rod and Piston-to-stroke Ratio

The rod-to-stroke ratio is the ratio of the length of the connecting rod to the length of the piston stroke. A longer rod reduces sidewise pressure of the piston on the cylinder wall and the stress forces, increasing engine life. It also increases the cost and engine height and weight.

A "square engine" is an engine with a bore diameter equal to its stroke length. An engine where the bore diameter is larger than its stroke length is an oversquare engine, conversely, an engine with a bore diameter that is smaller than its stroke length is an undersquare engine.

Valve Train

The valves are typically operated by a camshaft rotating at half the speed of the crankshaft. It has a series of cams along its length, each designed to open a valve during the appropriate part of an intake or exhaust stroke. A tappet between valve and cam is a contact surface on which the cam slides to open the valve. Many engines use one or more camshafts "above" a row (or each row) of cylinders, as in the illustration, in which each cam directly actuates a valve through a flat tappet. In other engine designs the camshaft is in the crankcase, in which case each cam usually contacts a push rod, which contacts a rocker arm that opens a valve, or in case of a flathead engine a push rod is not necessary. The overhead cam design typically allows higher engine speeds because it provides the most direct path between cam and valve.

Valve Clearance

Valve clearance refers to the small gap between a valve lifter and a valve stem that ensures that the valve completely closes. On engines with mechanical valve adjustment, excessive clearance causes noise from the valve train. A too small valve clearance can result in the valves not closing properly, this results in a loss of performance and possibly overheating of exhaust valves. Typically, the clearance must be readjusted each 20,000 miles (32,000 km) with a feeler gauge.

Most modern production engines use hydraulic lifters to automatically compensate for valve train component wear. Dirty engine oil may cause lifter failure.

Energy Balance

Otto engines are about 30% efficient; in other words, 30% of the energy generated by combustion is converted into useful rotational energy at the output shaft of the engine, while the remainder being losses due to waste heat, friction and engine accessories. There are a number of ways to recover some of the energy lost to waste heat. The use of a Turbocharger in Diesel engines is very effective by boosting incoming air pressure and in effect provides the same increase in performance as having more displacement. The Mack Truck company, decades ago, developed a turbine system that converted waste heat into kinetic energy that it fed back into the engine's transmission. In 2005, BMW announced the development of the turbosteamer, a two-stage heat-recovery system similar to the Mack system that recovers 80% of the energy in the exhaust gas and raises the efficiency of an Otto engine by 15%. By contrast, a six-stroke engine may reduce fuel consumption by as much as 40%.

Modern engines are often intentionally built to be slightly less efficient than they could otherwise be. This is necessary for emission controls such as exhaust gas recirculation and catalytic converters that reduce smog and other atmospheric pollutants. Reductions in efficiency may be counteracted with an engine control unit using lean burn techniques.

In the United States, the Corporate Average Fuel Economy mandates that vehicles must achieve an average of 34.9 mpg_{-US} (6.7 L/100 km; 41.9 mpg_{-imp}) compared to the current standard of 25 mpg_{-US} (9.4 L/100 km; 30.0 mpg_{-imp}). As automakers look to meet these standards by 2016, new ways of engineering the traditional internal combustion engine (ICE) have to be considered. Some potential solutions to increase fuel efficiency to meet new mandates include firing after the piston is farthest from the crankshaft, known as top dead centre, and applying the Miller cycle. Together, this redesign could significantly reduce fuel consumption and NOx emissions.

Interference Engine

An interference engine is a type of 4-stroke internal combustion piston engine in which one or more valves in the fully open position extends into any area that the piston may travel into. By contrast, in a non-interference engine the piston does not travel into any area into which the valves open. Interference engines rely on timing gears, chains, or belts to prevent the piston from striking the valves by ensuring that the valves are closed when the piston is near top dead center. Interference engines are prevalent among modern production automobiles and many other 4-stroke engine applications; the main advantage is that it allows engine designers to maximize the engine's compression ratio. However, such engines risk major internal damage if the piston strikes the valve(s) due to failure or poor maintenance of the timing components and/or camshaft(s).

Timing Gear Failure

A pair of valves bent by collision with a piston after timing belt breakage. The engine was running at 4500 RPM.

In interference engine designs, regular belt or chain service is especially important as incorrect timing may result in the pistons and valves colliding and causing extensive engine damage and therefore costly repairs. The piston will likely bend the valves or if a piece of valve or piston is broken off within the cylinder, the broken piece may cause severe damage within the cylinder, possibly affecting the connecting rods.

Belt Versus Chain

Many manufacturers who were using belts for valve timing have gone back to using chains on new engine offerings, especially on interference designs. During the peak popularity of the belt, chains or cogwheels were used almost exclusively on overhead valve (OHV) engines (which rarely are equipped with belts, regardless of the manufacturer and time of design) and almost all overhead camshaft (OHC) engines received belts. However, chains are lately becoming more popular for OHC designs.

Intake valves bent during a timing belt failure incident.

Belt Advantages

Some non-interference designs have retained belts due to the risk of engine damage from a belt failure being minimal. Some manufacturers liked the belt's quietness compared to the chain, and the ability to make additional profits from routine belt service.

Chain Advantages

Chains, in many cases, last the life of the engine, rarely requiring maintenance and helping to lower the cost of ownership for car buyers who are conscious of that statistic. Also, it was discoveredthat the sound difference between the two was negligible.

Two-stroke Engine

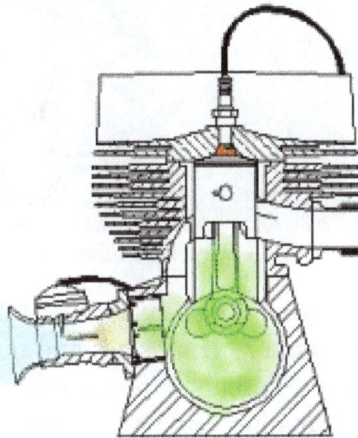

A two-stroke, or two-cycle, engine is a type of internal combustion engine which completes a power cycle with two strokes (up and down movements) of the piston during only one crankshaft revolution. This is in contrast to a "four-stroke engine", which requires four strokes of the piston to complete a power cycle. In a two-stroke engine, the end of the combustion stroke and the beginning of the compression stroke happen simultaneously, with the intake and exhaust (or scavenging) functions occurring at the same time.

Two-stroke engines often have a high power-to-weight ratio, power being available in a narrow range of rotational speeds called the "power band". Compared to four-stroke engines, two-stroke engines have a greatly reduced number of moving parts, and so can be more compact and significantly lighter.

The first commercial two-stroke engine involving in-cylinder compression is attributed to Scottish engineer Dugald Clerk, who patented his design in 1881. However, unlike most later two-stroke engines, his had a separate charging cylinder. The crankcase-scavenged engine, employing the area below the piston as a charging pump, is generally credited to Englishman Joseph Day. The first truly practical two-stroke engine is attributed to Yorkshireman Alfred Angas Scott, who started producing twin-cylinder water-cooled motorcycles in 1908.

Gasoline (spark ignition) versions are particularly useful in lightweight or portable applications such as chainsaws and motorcycles. However, when weight and size are not an issue, the cycle's potential for high thermodynamic efficiency makes it ideal for diesel compression ignition engines operating in large, weight-insensitive applications, such as marine propulsion, railway locomotives and electricity generation. In a two-stroke engine, the heat transfer from the engine to the cooling system is less than in a four-stroke, which means that two-stroke engines can be more efficient. However, crankcase-compression two-stroke engines, such as common small gasoline-powered engines, create more exhaust emissions than four-stroke engines because their two-stroke oil (petroil) lubrication mixture is also burned in the engine, due to the engine's total-loss oiling system.

Applications

A two-stroke minibike

The two-stroke petrol (gasoline) engine was very popular throughout the 20th century in motorcycles and small-engined devices, such as chainsaws and outboard motors. They were also used in some cars such as the Saab 93 and Trabant, a few tractors and many ships. Part of their appeal was their simple design (and resulting low cost) and often high power-to-weight ratio. The lower cost to rebuild and maintain made the two stroke engine very popular, until for the USA their EPA mandated more stringent emission controls in 1978 (taking effect in 1980) and in 2004 (taking effect in 2005 and 2010). The industry largely responded by switching to four-stroke petrol engines, which emit less pollution. Most small designs use petroil (two-stroke oil)) lubrication, with the oil being burned in the combustion chamber, causing "blue smoke" and other types of exhaust pollution. This is a major reason why two-stroke engines were replaced by four-stroke engines in many applications.

Lateral view of a two-stroke Forty series British Seagull outboard engine, the serial number dates it to 1954/1955

Simple two-stroke petrol engines continue to be commonly used in high-power, handheld applications such as string trimmers and chainsaws. The light weight, and light-weight spinning parts give important operational and safety advantages. For example, a four-stroke engine to power a chainsaw operating in any position would be much more expensive and complex than a two-stroke engine that uses a gasoline-oil mixture.

These engines are preferred for small, portable, or specialized machine applications such as outboard motors, high-performance, small-capacity motorcycles, mopeds, underbones, scooters, tuk-tuks, snowmobiles, karts, ultralights, model airplanes (and other model vehicles), lawnmowers, chainsaws, weed-wackers and dirt bikes.

The two-stroke cycle is also used in many diesel engines, most notably large industrial and marine engines, as well as some trucks and heavy machinery.

A number of mainstream automobile manufacturers have used two-stroke engines in the past, including the Swedish Saab and German manufacturers DKW, Auto-Union, VEB Sachsenring Automobilwerke Zwickau, and VEB Automobilwerk Eisenach. The Japanese manufacturer Suzuki did the same in the 1970s. Production of two-stroke cars ended in the 1980s in the West, but Eastern Bloc countries continued until around 1991, with the Trabant and Wartburg in East Germany. Lotus of Norfolk, UK, has a prototype direct-injection two-stroke engine intended for alcohol fuels called the Omnivore which it is demonstrating in a version of the Exige. As this uses direct fuel injection, there are dramatic decreases in emission levels and increases in fuel efficiency.

Different Two-stroke Design Types

Although the principles remain the same, the mechanical details of various two-stroke engines differ depending on the type. The design types vary according to the method of introducing the charge to the cylinder, the method of scavenging the cylinder (exchanging burnt exhaust for fresh mixture) and the method of exhausting the cylinder.

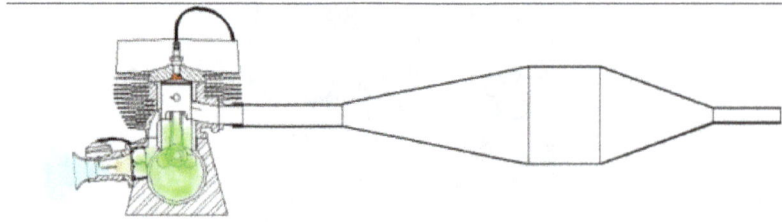

A two-stroke engine, in this case with an expansion chamber illustrates the effect of a reflected pressure wave on the fuel charge. This is important for maximum charge pressure (volumetric efficiency) and fuel economy. It is used on most high-performance engine designs.

Piston-controlled Inlet Port

Piston port is the simplest of the designs and the most common in small two-stroke engines. All functions are controlled solely by the piston covering and uncovering the ports as it moves up and down in the cylinder. In the 1970s, Yamaha worked out some basic principles for this system. They found that, in general, widening an exhaust port increases the power by the same amount as raising the port, but the power band does not narrow as it does when the port is raised. However, there is a mechanical limit to the width of a single exhaust port, at about 62% of the bore diameter for reasonable ring life. Beyond this, the rings will bulge into the exhaust port and wear quickly. A maximum is 70% of bore width is possible in racing engines, where rings are changed every few races. Intake duration is between 120 and 160 degrees. Transfer port time is set at a minimum of 26 degrees. The strong low pressure pulse of a racing two-stroke expansion chamber can drop the pressure to -7 PSI when the piston is at bottom dead center, and the transfer ports nearly wide open. One of the reasons for high fuel consumption in 2-strokes is that some of the incoming pressurized fuel/air mixture is forced across the top of the piston, where it has a cooling action, and straight out the exhaust pipe. An expansion chamber with a strong reverse pulse will stop this out-going flow. A fundamental difference from typical four-stroke engines is that the two-stroke's crankcase is sealed and forms part of the induction process in gasoline and hot bulb engines. Diesel two-strokes often add a Roots blower or piston pump for scavenging.

Reed Inlet Valve

The reed valve is a simple but highly effective form of check valve commonly fitted in the intake tract of the piston-controlled port. They allow asymmetric intake of the fuel charge, improving power and economy, while widening the power band. They are widely used in motorcycle, ATV and marine outboard engines.

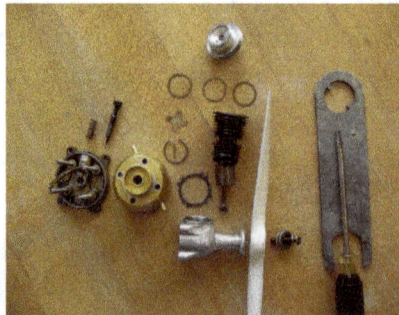

A Cox Babe Bee 0.049 cubic inch (0.8 cubic cm) reed valve engine, disassembled, uses glow plug ignition. The mass is 64 grams.

Rotary Inlet Valve

The intake pathway is opened and closed by a rotating member. A familiar type sometimes seen on small motorcycles is a slotted disk attached to the crankshaft which covers and uncovers an opening in the end of the crankcase, allowing charge to enter during one portion of the cycle.

Another form of rotary inlet valve used on two-stroke engines employs two cylindrical members with suitable cutouts arranged to rotate one within the other - the inlet pipe having passage to the crankcase only when the two cutouts coincide. The crankshaft itself may form one of the members, as in most glow plug model engines. In another embodiment, the crank disc is arranged to be a close-clearance fit in the crankcase, and is provided with a cutout which lines up with an inlet passage in the crankcase wall at the appropriate time, as in the Vespa motor scooter.

The advantage of a rotary valve is it enables the two-stroke engine's intake timing to be asymmetrical, which is not possible with piston port type engines. The piston port type engine's intake timing opens and closes before and after top dead center at the same crank angle, making it symmetrical, whereas the rotary valve allows the opening to begin earlier and close earlier.

Rotary valve engines can be tailored to deliver power over a wider speed range or higher power over a narrower speed range than either piston port or reed valve engine. Where a portion of the rotary valve is a portion of the crankcase itself, it is particularly important that no wear is allowed to take place.

Cross-flow-scavenged

In a cross-flow engine, the transfer and exhaust ports are on opposite sides of the cylinder, and a deflector on the top of the piston directs the fresh intake charge into the upper part of the cylinder, pushing the residual exhaust gas down the other side of the deflector and out the exhaust port. The deflector increases the piston's weight and exposed surface area, affecting piston cooling and also making it difficult to achieve an efficient combustion chamber shape. This design has been superseded since the 1960s by the loop scavenging method (below), especially for motorbikes, although for smaller or slower engines, such as lawn mowers, the cross-flow-scavenged design can be an acceptable approach.

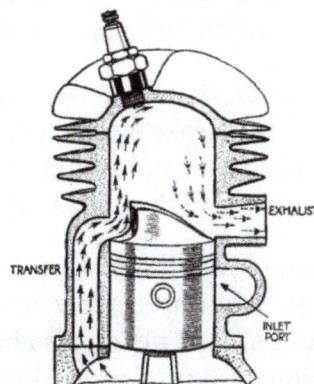

Deflector piston with cross-flow scavenging

Loop-scavenged

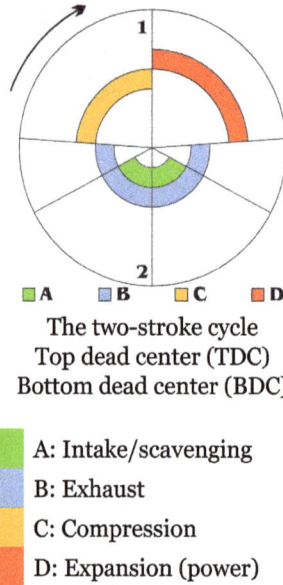

The two-stroke cycle
Top dead center (TDC)
Bottom dead center (BDC)

A: Intake/scavenging

B: Exhaust

C: Compression

D: Expansion (power)

This method of scavenging uses carefully shaped and positioned transfer ports to direct the flow of fresh mixture toward the combustion chamber as it enters the cylinder. The fuel/air mixture strikes the cylinder head, then follows the curvature of the combustion chamber, and then is deflected downward.

This not only prevents the fuel/air mixture from traveling directly out the exhaust port, but also creates a swirling turbulence which improves combustion efficiency, power and economy. Usually, a piston deflector is not required, so this approach has a distinct advantage over the cross-flow scheme (above).

Often referred to as "Schnuerle" (or "Schnürle") loop scavenging after the German inventor of an early form in the mid-1920s, it became widely adopted in that country during the 1930s and spread further afield after World War II.

Loop scavenging is the most common type of fuel/air mixture transfer used on modern two-stroke engines. Suzuki was one of the first manufacturers outside of Europe to adopt loop-scavenged two-stroke engines. This operational feature was used in conjunction with the expansion chamber exhaust developed by German motorcycle manufacturer, MZ and Walter Kaaden.

Loop scavenging, disc valves and expansion chambers worked in a highly coordinated way to significantly increase the power output of two-stroke engines, particularly from the Japanese manufacturers Suzuki, Yamaha and Kawasaki. Suzuki and Yamaha enjoyed success in grand Prix motorcycle racing in the 1960s due in no small way to the increased power afforded by loop scavenging.

An additional benefit of loop scavenging was the piston could be made nearly flat or slightly dome shaped, which allowed the piston to be appreciably lighter and stronger, and consequently to tolerate higher engine speeds. The "flat top" piston also has better thermal properties and is less prone to uneven heating, expansion, piston seizures, dimensional changes and compression losses.

SAAB built 750 and 850 cc 3-cylinder engines based on a DKW design that proved reasonably successful employing loop charging. The original SAAB 92 had a two-cylinder engine of comparatively low efficiency. At cruising speed, reflected wave exhaust port blocking occurred at too low a frequency. Using the asymmetric three-port exhaust manifold employed in the identical DKW engine improved fuel economy.

The 750 cc standard engine produced 36 to 42 hp, depending on the model year. The Monte Carlo Rally variant, 750 cc (with a filled crankshaft for higher base compression), generated 65 hp. An 850 cc version was available in the 1966 SAAB Sport (a standard trim model in comparison to the deluxe trim of the Monte Carlo). Base compression comprises a portion of the overall compression ratio of a two-stroke engine. Work published at SAE in 2012 points that loop scavenging is under every circumstance more efficient than cross-flow scavenging.

Uniflow-scavenged

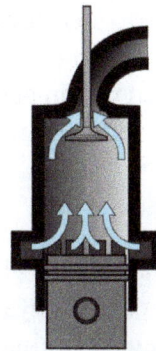

Uniflow scavenging

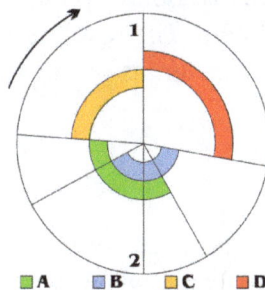

The uniflow two-stroke cycle
Top dead center (TDC)
Bottom dead center (BDC)

A: Intake (effective scavenging 135°–225°; necessarily symmetric about BDC; Diesel injection is usually initiated at 4° before TDC)

B: Exhaust

C: Compression

D: Expansion (power)

In a uniflow engine, the mixture, or "charge air" in the case of a diesel, enters at one end of the cylinder controlled by the piston and the exhaust exits at the other end controlled by an exhaust valve or piston. The scavenging gas-flow is therefore in one direction only, hence the name uniflow. The valved arrangement is common in on-road, off-road and stationary two-stroke engines (Detroit Diesel), certain small marine two-stroke engines (Gray Marine), certain railroad two-stroke diesel locomotives (Electro-Motive Diesel) and large marine two-stroke main propulsion engines (Wärtsilä). Ported types are represented by the opposed piston design in which there are two pistons in each cylinder, working in opposite directions such as the Junkers Jumo 205 and Napier Deltic. The once-popular split-single design falls into this class, being effectively a folded uniflow. With advanced angle exhaust timing, uniflow engines can be supercharged with a crankshaft-driven (piston or Roots) blower.

Stepped Piston Engine

The piston of this engine is "top-hat" shaped; the upper section forms the regular cylinder, and the lower section performs a scavenging function. The units run in pairs, with the lower half of one piston charging an adjacent combustion chamber.

This system is still partially dependent on total loss lubrication (for the upper part of the piston), the other parts being sump lubricated with cleanliness and reliability benefits. The piston weight is only about 20% heavier than a loop-scavenged piston because skirt thicknesses can be less. Bernard Hooper Engineering Ltd. (BHE) is one of the more recent engine developers using this approach.

Power Valve Systems

Many modern two-stroke engines employ a power valve system. The valves are normally in or around the exhaust ports. They work in one of two ways: either they alter the exhaust port by closing off the top part of the port, which alters port timing, such as Ski-doo R.A.V.E, Yamaha YPVS, Honda RC-Valve, Kawasaki K.I.P.S., Cagiva C.T.S. or Suzuki AETC systems, or by altering the volume of the exhaust, which changes the resonant frequency of the expansion chamber, such as the Suzuki SAEC and Honda V-TACS system. The result is an engine with better low-speed power without sacrificing high-speed power. However, as power valves are in the hot gas flow they need regular maintenance to perform well.

Direct Injection

Direct injection has considerable advantages in two-stroke engines, eliminating some of the waste and pollution caused by carbureted two-strokes where a proportion of the fuel/air mixture entering the cylinder goes directly out, unburned, through the exhaust port. Two systems are in use, low-pressure air-assisted injection, and high pressure injection.

Since the fuel does not pass through the crankcase, a separate source of lubrication is needed.

Two-stroke Diesel Engine

Diesel engines rely solely on the heat of compression for ignition. In the case of Schnuerle ported and loop-scavenged engines, intake and exhaust happens via piston-controlled ports. A uniflow diesel engine takes in air via scavenge ports, and exhaust gases exit through an overhead poppet valve. Two-stroke diesels are all scavenged by forced induction. Some designs use a mechanically driven Roots blower, whilst marine diesel engines normally use exhaust-driven turbochargers, with electrically driven auxiliary blowers for low-speed operation when exhaust turbochargers are unable to deliver enough air.

Brons two-stroke V8 Diesel engine driving a N.V. Heemaf generator.

Marine two-stroke diesel engines directly coupled to the propeller are able to start and run in either direction as required. The fuel injection and valve timing is mechanically readjusted by using a different set of cams on the camshaft. Thus, the engine can be run in reverse to move the vessel backwards.

Lubrication

Most small petrol two-stroke engines cannot be lubricated by oil contained in their crankcase and sump, since the crankcase is already being used to pump fuel-air mixture into the cylinder. Traditionally, the moving parts (both rotating crankshaft and sliding piston) were lubricated by a premixed fuel-oil mixture (at a ratio between 16:1 and 100:1). As late as the 1970s, petrol stations would often have a separate pump to deliver such a premix fuel to motorcycles. Even then, in many cases, the rider would carry a bottle of their own two-stroke oil.

Two-stroke oils which became available worldwide in the 1970s are specifically designed to mix with petrol and be burnt in the combustion chamber without leaving undue unburnt oil or ash. This led to a marked reduction in spark plug fouling, which had previously been a factor in two-stroke engines.

More recent two-stroke engines might pump lubrication from a separate tank of two-stroke oil. The supply of this oil is controlled by the throttle position. An example machine is Yamaha's PW80 (Pee-wee), a small, 80cc two-stroke dirt bike designed for young children. The technology is referred to as auto-lube. This is still a total-loss system with the oil being burnt the same as in the pre-mix system; however, given that the oil is not properly mixed with the fuel when burned in the combustion chamber, it translates into a slightly more efficient lubrication. This lubrication method also pays dividends in terms of user friendliness by eliminating the user's need to mix the gasoline at every refill, makes the motor much less susceptible to atmospheric conditions (Ambient temperature, elevation) and ensures a constant, unvarying fuel-oil ratio in the combustion chamber. However, this method of lubrication is still ultimately the same as premixed gasoline in that the oil is still burnt (albeit not as completely as pre-mix) and the gas is still mixed with the oil, although not as thoroughly as in pre-mix. In addition, this method requires extra mechanical parts to pump the oil from the separate tank, to the carburetor or throttle body. In applications where performance, simplicity and/or dry weight are significant considerations, the pre-mix lubrication method is almost always used. For example, a two-stroke engine in a motocross bike pays major consideration to performance, simplicity and weight. Chainsaws and brush cutters must be as light as possible to reduce user fatigue and hazard, especially when used in a professional work environment.

All two-stroke engines running on a petrol/oil mix will suffer oil starvation if forced to rotate at speed with the throttle closed, e.g. motorcycles descending long hills and perhaps when decelerating gradually from high speed by changing down through the gears. Two-stroke cars (such as those that were popular in Eastern Europe in the mid-20th century) were in particular danger and were usually fitted with freewheel mechanisms in the powertrain, allowing the engine to idle when the throttle was closed, requiring the use of the brakes in all slowing situations.

Large two-stroke engines, including diesels, normally use a sump lubrication system similar to four-stroke engines. The cylinder must still be pressurized, but this is not done from the crankcase, but by an ancillary Roots-type blower or a specialized turbocharger (usually a turbo-compressor system) which has a "locked" compressor for starting (and during which it is powered by the engine's crankshaft), but which is "unlocked" for running (and during which it is powered by the engine's exhaust gases flowing through the turbine).

Two-stroke Reversibility

For the purpose of this discussion, it is convenient to think in motorcycle terms, where the exhaust pipe faces into the cooling air stream, and the crankshaft commonly spins in the same axis and direction as do the wheels i.e. "forward". Some of the considerations discussed here apply to four-stroke engines (which cannot reverse their direction of rotation without considerable modification), almost all of which spin forward, too.

Regular gasoline two-stroke engines will run backwards for short periods and under light load with little problem, and this has been used to provide a reversing facility in microcars, such as the Messerschmitt KR200, that lacked reverse gearing. Where the vehicle has electric starting, the motor will be turned off and restarted backwards by turning the key in the opposite direction. Two-stroke golf carts have used a similar kind of system. Traditional flywheel magnetos (using contact-breaker points, but no external coil) worked equally well in reverse because the cam controlling the points is symmetrical, breaking contact before top dead center (TDC) equally well whether running forwards or backwards. Reed-valve engines will run backwards just as well as piston-controlled porting, though rotary valve engines have asymmetrical inlet timing and will not run very well.

There are serious disadvantages to running many engines backwards under load for any length of time, and some of these reasons are general, applying equally to both two-stroke and four-stroke engines. This disadvantage is accepted in most cases where cost, weight and size are major considerations. The problem comes about because in "forwards" running the major thrust face of the piston is on the back face of the cylinder which, in a two-stroke particularly, is the coolest and best-lubricated part. The forward face of the piston in a trunk engine is less well-suited to be the major thrust face since it covers and uncovers the exhaust port in the cylinder, the hottest part of the engine, where piston lubrication is at its most marginal. The front face of the piston is also more vulnerable since the exhaust port, the largest in the engine, is in the front wall of the cylinder. Piston skirts and rings risk being extruded into this port, so it is always better to have them pressing hardest on the opposite wall (where there are only the transfer ports in a crossflow engine) and there is good support. In some engines, the small end is offset to reduce thrust in the intended rotational direction and the forward face of the piston has been made thinner and lighter to compensate - but when running backwards, this weaker forward face suffers increased mechanical stress it was not designed to resist. This can be avoided by the use of crossheads and also using thrust bearings to isolate the engine from end loads.

Large two-stroke ship diesels are sometimes made to be reversible. Like four-stroke ship engines (some of which are also reversible) they use mechanically operated valves, so require additional camshaft mechanisms. These engine use crossheads to eliminate sidethrust on the piston and isolate the under-piston space from the crankcase.

On top of other considerations, the oil-pump of a modern two-stroke may not work in reverse, in which case the engine will suffer oil starvation within a short time. Running a motorcycle engine backwards is relatively easy to initiate, and in rare cases, can be triggered by a back-fire. It is not advisable.

Model airplane engines with reed-valves can be mounted in either tractor or pusher configuration without needing to change the propeller. These motors are compression ignition, so there are no ignition timing issues and little difference between running forward and running backward.

Gas Turbine

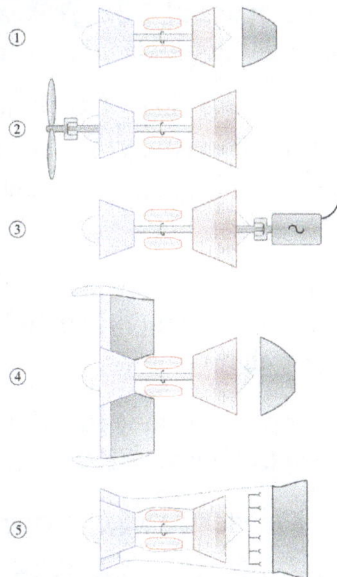

Examples of gas turbine configurations: (1) turbojet, (2) turboprop, (3) turboshaft (electric generator), (4) high-bypass turbofan, (5) low-bypass afterburning turbofan

A gas turbine, also called a combustion turbine, is a type of internal combustion engine. It has an upstream rotating compressor coupled to a downstream turbine, and a combustion chamber in between.

The basic operation of the gas turbine is similar to that of the steam power plant except that air is used instead of water. Fresh atmospheric air flows through a compressor that brings it to higher pressure. Energy is then added by spraying fuel into the air and igniting it so the combustion generates a high-temperature flow. This high-temperature high-pressure gas enters a turbine, where it expands down to the exhaust pressure, producing a shaft work output in the process. The turbine shaft work is used to drive the compressor and other devices such as an electric generator that may be coupled to the shaft. The energy that is not used for shaft work comes out in the exhaust gases, so these have either a high temperature or a high velocity. The purpose of the gas turbine determines the design so that the most desirable energy form is maximized. Gas turbines are used to power aircraft, trains, ships, electrical generators, and tanks.

History

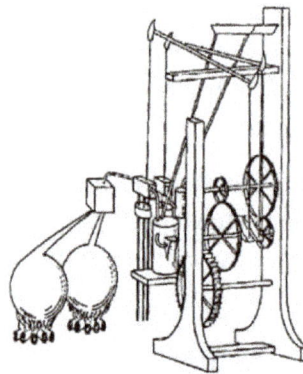

Sketch of John Barber's gas turbine, from his patent

- 50: Hero's Engine (*aeolipile*) — Apparently, Hero's steam engine was taken to be no more than a toy, and thus its full potential not realized for centuries.

- 1000: The "Trotting Horse Lamp" (Chinese: 走马灯) was used by the Chinese at lantern fairs as early as the Northern Song dynasty. When the lamp is lit, the heated airflow rises and drives an impeller with horse-riding figures attached on it, whose shadows are then projected onto the outer screen of the lantern.

- 1500: The "Chimney Jack" was drawn by Leonardo da Vinci: Hot air from a fire rises through a single-stage axial turbine rotor mounted in the exhaust duct of the fireplace and turning the roasting spit by gear/ chain connection.

- 1629: Jets of steam rotated an impulse turbine that then drove a working stamping mill by means of a bevel gear, developed by Giovanni Branca.

- 1678: Ferdinand Verbiest built a model carriage relying on a steam jet for power.

- 1791: A patent was given to John Barber, an Englishman, for the first true gas turbine. His invention had most of the elements present in the modern day gas turbines. The turbine was designed to power a horseless carriage.

- 1861: British patent no. 1633 was granted to Marc Antoine Francois Mennons for a "Caloric engine". The patent shows that it was a gas turbine and the drawings show it applied to a locomotive. Also named in the patent was Nicolas de Telescheff (otherwise Nicholas A. Teleshov), a Russian aviation pioneer.

- 1872: A gas turbine engine was designed by Franz Stolze, but the engine never ran under its own power.

- 1894: Sir Charles Parsons patented the idea of propelling a ship with a steam turbine, and built a demonstration vessel, the *Turbinia*, easily the fastest vessel afloat at the time. This principle of propulsion is still of some use.

- 1895: Three 4-ton 100 kW Parsons radial flow generators were installed in Cambridge Power Station, and used to power the first electric street lighting scheme in the city.

- 1899: Charles Gordon Curtis patented the first gas turbine engine in the USA ("Apparatus for generating mechanical power", Patent No. US635,919).

- 1900: Sanford Alexander Moss submitted a thesis on gas turbines. In 1903, Moss became an engineer for General Electric's Steam Turbine Department in Lynn, Massachusetts. While there, he applied some of his concepts in the development of the turbosupercharger. His design used a small turbine wheel, driven by exhaust gases, to turn a supercharger.

- 1903: A Norwegian, Ægidius Elling, was able to build the first gas turbine that was able to produce more power than needed to run its own components, which was considered an achievement in a time when knowledge about aerodynamics was limited. Using rotary compressors and turbines it produced 11 hp (massive for those days).

- 1906: The Armengaud-Lemale turbine engine in France with water-cooled combustion chamber.

- 1910: Holzwarth impulse turbine (pulse combustion) achieved 150 kilowatts.

- 1913: Nikola Tesla patents the Tesla turbine based on the boundary layer effect.

- 1920s The practical theory of gas flow through passages was developed into the more formal (and applicable to turbines) theory of gas flow past airfoils by A. A. Griffith resulting in the publishing in 1926 of *An Aerodynamic Theory of Turbine Design*. Working testbed designs of axial turbines suitable for driving a propellor were developed by the Royal Aeronautical Establishment proving the efficiency of aerodynamic shaping of the blades in 1929.

- 1930: Having found no interest from the RAF for his idea, Frank Whittle patented the design for a centrifugal gas turbine for jet propulsion. The first successful use of his engine was in April 1937.

- 1932: BBC Brown, Boveri & Cie of Switzerland starts selling axial compressor and turbine turbosets as part of the turbocharged steam generating Velox boiler. Following the gas turbine principle, the steam evaporation tubes are arranged within the gas turbine combustion chamber; the first Velox plant was erected in Mondeville, France.

- 1934: Raúl Pateras de Pescara patented the free-piston engine as a gas generator for gas turbines.

- 1936: Hans von Ohain and Max Hahn in Germany were developing their own patented engine design.

- 1936 Whittle with others backed by investment forms Power Jets Ltd

- 1937 The first Power Jets engine runs, and impresses Henry Tizard such that he secures government funding for its further development.

- 1939: First 4 MW utility power generation gas turbine from BBC Brown, Boveri & Cie. for an emergency power station in Neuchâtel, Switzerland.

- 1946 National Gas Turbine Establishment formed from Power Jets and the RAE turbine division bring together Whittle and Hayne Constant's work. In Beznau, Switzerland the first commercial reheated/recuperated unit generating 27 MW was commissioned.

- 1963 Pratt and Whitney introduce the GG4/FT4 which is the first commercial aeroderivative gas turbine.

- 2011 Mitsubishi Heavy Industries tests the first >60% efficiency gas turbine (the M501J) at its Takasago works.

Theory of Operation

In an ideal gas turbine, gases undergo three thermodynamic processes: an isentropic compression, an isobaric (constant pressure) combustion and an isentropic expansion. Together, these make up the Brayton cycle.

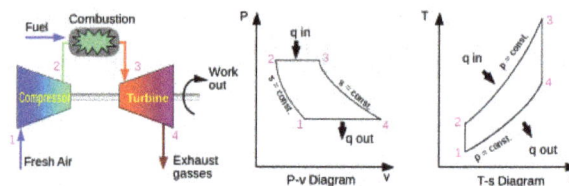

Brayton cycle

In a real gas turbine, mechanical energy is changed irreversibly (due to internal friction and turbulence) into pressure and thermal energy when the gas is compressed (in either a centrifugal or axial compressor). Heat is added in the combustion chamber and the specific volume of the gas increases, accompanied by a slight loss in pressure. During expansion through the stator and rotor passages in the turbine, irreversible energy transformation once again occurs.

If the engine has a power turbine added to drive an industrial generator or a helicopter rotor, the exit pressure will be as close to the entry pressure as possible with only enough energy left to overcome the pressure losses in the exhaust ducting and expel the exhaust. For a turboprop engine there will be a particular balance between propeller power and jet thrust which gives the most economical operation. In a jet engine only enough pressure and energy is extracted from the flow to drive the compressor and other components. The remaining high pressure gases are accelerated to provide a jet to propel an aircraft.

The smaller the engine, the higher the rotation rate of the shaft(s) must be to attain the required blade tip speed. Blade-tip speed determines the maximum pressure ratios that can be obtained by the turbine and the compressor. This, in turn, limits the maximum power and efficiency that can be obtained by the engine. In order for tip speed to remain constant, if the diameter of a rotor is reduced by half, the rotational speed must double. For example, large jet engines operate around 10,000 rpm, while micro turbines spin as fast as 500,000 rpm.

Mechanically, gas turbines can be considerably less complex than internal combustion piston engines. Simple turbines might have one main moving part, the compressor/shaft/turbine rotor assembly (see image above), with other moving parts in the fuel system. However, the precision manufacture required for components and the temperature resistant alloys necessary for high efficiency often make the construction of a simple gas turbine more complicated than a piston engine.

More advanced gas turbines (such as those found in modern jet engines) may have 2 or 3 shafts (spools), hundreds of compressor and turbine blades, movable stator blades, and extensive external tubing for fuel, oil and air systems.

Thrust bearings and journal bearings are a critical part of design. Traditionally, they have been hydrodynamic oil bearings, or oil-cooled ball bearings. These bearings are being surpassed by foil bearings, which have been successfully used in micro turbines and auxiliary power units.

Creep

A major challenge facing turbine design is reducing the creep that is induced by the high temperatures. Because of the stresses of operation, turbine materials become damaged through these mechanisms. As temperatures are increased in an effort to improve turbine efficiency, creep becomes more significant. To limit creep, thermal coatings and superalloys with solid-solution strengthening and grain boundary strengthening are used in blade designs. Protective coatings are used in to reduce the thermal damage and to limit oxidation. These coatings are often stabilized zirconium dioxide-based ceramics. Using a thermal protective coating limits the temperature exposure of the nickel superalloy. This reduces the creep mechanisms experienced in the blade. Oxidation coatings limit efficiency losses caused by a buildup on the outside of the blades, which is especially important in the high-temperature environment. The nickel-based blades are alloyed with aluminum and titanium to improve strength and creep resistance. The microstructure of these alloys is composed of different regions of composition. A uniform dispersion of the gamma-prime phase – a combination of nickel, aluminum, and titanium – promotes the strength and creep resistance of the blade due to the microstructure. Refractory elements such as rhenium and ruthenium can be added to the alloy to improve creep strength. The addition of these elements reduces the diffusion of the gamma prime phase, thus preserving the fatigue resistance, strength, and creep resistance.

Types

Jet Engines

Airbreathing jet engines are gas turbines optimized to produce thrust from the exhaust gases, or from ducted fans connected to the gas turbines. Jet engines that produce thrust from the direct impulse of exhaust gases are often called turbojets, whereas those that generate thrust with the addition of a ducted fan are often called turbofans or (rarely) fan-jets.

typical axial-flow gas turbine turbojet, the J85, sectioned for display. Flow is left to right, multistage compressor on left, combustion chambers center, two-stage turbine on right

Gas turbines are also used in many liquid propellant rockets, the gas turbines are used to power a turbopump to permit the use of lightweight, low pressure tanks, which reduce the empty weight of the rocket.

Turboprop Engines

A turboprop engine is a turbine engine which drives an aircraft propeller using a reduction gear. Turboprop engines are used on small aircraft such as the general-aviation Cessna 208 Caravan and Embraer EMB 312 Tucano military trainer, medium-sized commuter aircraft such as the Bombardier Dash 8 and large aircraft such as the Airbus A400M transport and the 60 year-old Tupolev Tu-95 strategic bomber.

Aeroderivative Gas Turbines

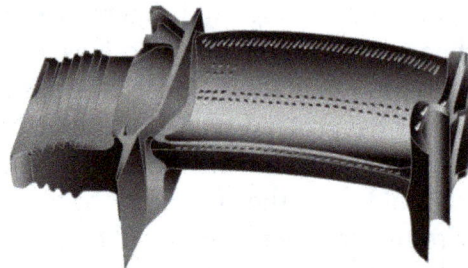

Diagram of a high-pressure film cooled turbine blade

Aeroderivatives are also used in electrical power generation due to their ability to be shut down, and handle load changes more quickly than industrial machines. They are also used in the marine industry to reduce weight. The General Electric LM2500, General Electric LM6000, Rolls-Royce RB211 and Rolls-Royce Avon are common models of this type of machine.

Amateur Gas Turbines

Increasing numbers of gas turbines are being used or even constructed by amateurs.

In its most straightforward form, these are commercial turbines acquired through military surplus or scrapyard sales, then operated for display as part of the hobby of engine collecting. In its most extreme form, amateurs have even rebuilt engines beyond professional repair and then used them to compete for the Land Speed Record.

The simplest form of self-constructed gas turbine employs an automotive turbocharger as the core component. A combustion chamber is fabricated and plumbed between the compressor and turbine sections.

More sophisticated turbojets are also built, where their thrust and light weight are sufficient to power large model aircraft. The Schreckling design constructs the entire engine from raw materials, including the fabrication of a centrifugal compressor wheel from plywood, epoxy and wrapped carbon fibre strands.

Several small companies now manufacture small turbines and parts for the amateur. Most turbojet-powered model aircraft are now using these commercial and semi-commercial microturbines, rather than a Schreckling-like home-build.

Auxiliary Power Units

APUs are small gas turbines designed to supply auxiliary power to larger, mobile, machines such as an aircraft. They supply:

- compressed air for air conditioning and ventilation,

- compressed air start-up power for larger jet engines,

- mechanical (shaft) power to a gearbox to drive shafted accessories or to start large jet engines, and

- electrical, hydraulic and other power-transmission sources to consuming devices remote from the APU.

Industrial Gas Turbines for Power Generation

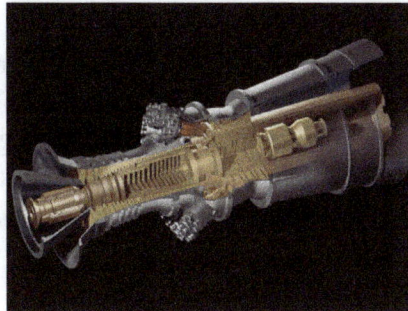

GE H series power generation gas turbine: in combined cycle configuration, this 480-megawatt unit has a rated thermal efficiency of 60%

Industrial gas turbines differ from aeronautical designs in that the frames, bearings, and blading are of heavier construction. They are also much more closely integrated with the devices they power— often an electric generator—and the secondary-energy equipment that is used to recover residual energy (largely heat).

They range in size from portable mobile plants to large, complex systems weighing more than a hundred tonnes housed in purpose-built buildings. When the gas turbine is used solely for shaft power, its thermal efficiency is about 30%. However, it may be cheaper to buy electricity than to generate it. Therefore, many engines are used in CHP (Combined Heat and Power) configurations that can be small enough to be integrated into portable container configurations.

Gas turbines can be particularly efficient when waste heat from the turbine is recovered by a heat recovery steam generator to power a conventional steam turbine in a combined cycle configuration. The 605 MW General Electric 9HA achieved a 62.22% efficiency rate with temperatures as high as 1,540 °C (2,800 °F). Aeroderivative gas turbines can also be used in combined cycles, leading to a higher efficiency, but it will not be as high as a specifically designed industrial gas turbine. They can also be run in a cogeneration configuration: the exhaust is used for space or water heating, or drives an absorption chiller for cooling the inlet air and increase the power output, technology known as Turbine Inlet Air Cooling.

Another significant advantage is their ability to be turned on and off within minutes, supplying power during peak, or unscheduled, demand. Since single cycle (gas turbine only) power plants are less efficient than combined cycle plants, they are usually used as peaking power plants, which operate anywhere from several hours per day to a few dozen hours per year—depending on the electricity demand and the generating capacity of the region. In areas with a shortage of base-load and load following power plant capacity or with low fuel costs, a gas turbine powerplant may regularly operate most hours of the day. A large single-cycle gas turbine typically produces 100 to 400 megawatts of electric power and has 35–40% thermal efficiency.

Industrial Gas Turbines for Mechanical Drive

Industrial gas turbines that are used solely for mechanical drive or used in collaboration with a recovery steam generator differ from power generating sets in that they are often smaller and feature a dual shaft design as opposed to single shaft. The power range varies from 1 megawatt up to 50 megawatts. These engines are connected directly or via a gearbox to either a pump or compressor assembly. The majority of installations are used within the oil and gas industries. Mechanical drive applications increase efficiency by around 2%.

Oil and Gas platforms require these engines to drive compressors to inject gas into the wells to force oil up via another bore, or to compress the gas for transportation. They're also often used to provide power for the platform. These platforms don't need to use the engine in collaboration with a CHP system due to getting the gas at an extremely reduced cost (often free from burn off gas). The same companies use pump sets to drive the fluids to land and across pipelines in various intervals.

Compressed Air Energy Storage

One modern development seeks to improve efficiency in another way, by separating the compressor and the turbine with a compressed air store. In a conventional turbine, up to half the generated power is used driving the compressor. In a compressed air energy storage configuration, power, perhaps from a wind farm or bought on the open market at a time of low demand and low price, is used to drive the compressor, and the compressed air released to operate the turbine when required.

Turboshaft Engines

Turboshaft engines are often used to drive compression trains (for example in gas pumping stations or natural gas liquefaction plants) and are used to power almost all modern helicopters. The primary shaft bears the compressor and the high speed turbine (often referred to as the *Gas Generator*), while a second shaft bears the low-speed turbine (a *power turbine* or *free-wheeling turbine* on helicopters, especially, because the gas generator turbine spins separately from the power turbine). In effect the separation of the gas generator, by a fluid coupling (the hot energy-rich combustion gases), from the power turbine is analogous to an automotive transmission's fluid coupling. This arrangement is used to increase power-output flexibility with associated highly-reliable control mechanisms.

Radial Gas Turbines

In 1963, Jan Mowill initiated the development at Kongsberg Våpenfabrikk in Norway. Various successors have made good progress in the refinement of this mechanism. Owing to a configuration that keeps heat away from certain bearings the durability of the machine is improved while the radial turbine is well matched in speed requirement.

Scale Jet Engines

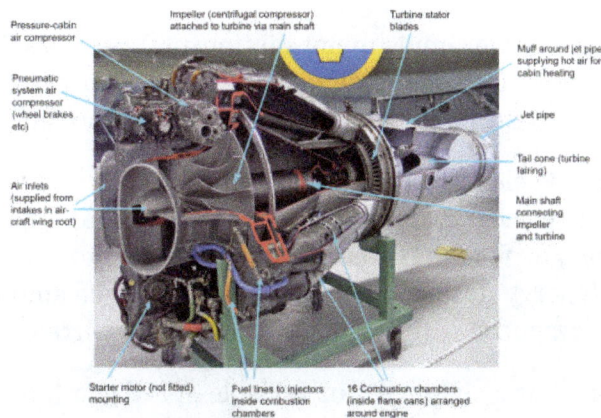

Scale jet engines are scaled down versions of this early full scale engine

Also known as miniature gas turbines or micro-jets.

With this in mind the pioneer of modern Micro-Jets, Kurt Schreckling, produced one of the world's first Micro-Turbines, the FD3/67. This engine can produce up to 22 newtons of thrust, and can be built by most mechanically minded people with basic engineering tools, such as a metal lathe.

Microturbines

Also known as:

- Turbo alternators

- Turbogenerator

Microturbines are becoming widespread in distributed power and combined heat and power applications, and are very promising for powering hybrid electric vehicles. They range from hand held units producing less than a kilowatt, to commercial sized systems that produce tens or hundreds of kilowatts. Basic principles of microturbine are based on micro-combustion.

Part of their claimed success is said to be due to advances in electronics, which allows unattended operation and interfacing with the commercial power grid. Electronic power switching technology eliminates the need for the generator to be synchronized with the power grid. This allows the generator to be integrated with the turbine shaft, and to double as the starter motor.

Microturbine systems have many claimed advantages over reciprocating engine generators, such as higher power-to-weight ratio, low emissions and few, or just one, moving part. Advantages are that microturbines may be designed with foil bearings and air-cooling operating without lubricating oil, coolants or other hazardous materials. Nevertheless, reciprocating engines overall are still cheaper when all factors are considered.Microturbines also have a further advantage of having the majority of the waste heat contained in the relatively high temperature exhaust making it simpler to capture, whereas the waste heat of reciprocating engines is split between its exhaust and cooling system.

However, reciprocating engine generators are quicker to respond to changes in output power requirement and are usually slightly more efficient, although the efficiency of microturbines is increasing. Microturbines also lose more efficiency at low power levels than reciprocating engines.

Reciprocating engines typically use simple motor oil (journal) bearings. Full-size gas turbines often use ball bearings. The 1000 °C temperatures and high speeds of microturbines make oil lubrication and ball bearings impractical; they require air bearings or possibly magnetic bearings.

When used in extended range electric vehicles the static efficiency drawback is irrelevant, since the gas turbine can be run at or near maximum power, driving an alternator to produce electricity either for the wheel motors, or for the batteries, as appropriate to speed and battery state. The batteries act as a "buffer" (energy storage) in delivering the required amount of power to the wheel motors, rendering throttle response of the gas turbine completely irrelevant.

There is, moreover, no need for a significant or variable-speed gearbox; turning an alternator at comparatively high speeds allows for a smaller and lighter alternator than would otherwise be the case. The superior power-to-weight ratio of the gas turbine and its fixed speed gearbox, allows for a much lighter prime mover than those in such hybrids as the Toyota Prius (which utilised a 1.8 litre petrol engine) or the Chevrolet Volt (which utilises a 1.4 litre petrol engine). This in turn allows a heavier weight of batteries to be carried, which allows for a longer electric-only range. Alternatively, the vehicle can use heavier types of batteries such as lead acid batteries (which are cheaper to buy) or safer types of batteries such as Lithium-Iron-Phosphate.

When gas turbines are used in extended-range electric vehicles, like those planned by Land-Rover/Range-Rover in conjunction with Bladon, or by Jaguar also in partnership with Bladon, the very poor throttling response (their high moment of rotational inertia) does not matter, because the gas turbine, which may be spinning at 100,000 rpm, is not directly, mechanically connected to the wheels. It was this poor throttling response that so bedevilled the 1950 Rover gas turbine-powered prototype motor car, which did not have the advantage of an intermediate electric drive train to provide sudden power spikes when demanded by the driver.

Gas turbines accept most commercial fuels, such as petrol, natural gas, propane, diesel, and kerosene as well as renewable fuels such as E85, biodiesel and biogas. However, when running on kerosene or diesel, starting sometimes requires the assistance of a more volatile product such as propane gas - although the new kero-start technology can allow even microturbines fuelled on kerosene to start without propane.

Microturbine designs usually consist of a single stage radial compressor, a single stage radial turbine and a recuperator. Recuperators are difficult to design and manufacture because they operate under high pressure and temperature differentials. Exhaust heat can be used for water heating, space heating, drying processes or absorption chillers, which create cold for air conditioning from heat energy instead of electric energy.

Typical microturbine efficiencies are 25 to 35%. When in a combined heat and power cogeneration system, efficiencies of greater than 80% are commonly achieved.

MIT started its millimeter size turbine engine project in the middle of the 1990s when Professor of Aeronautics and Astronautics Alan H. Epstein considered the possibility of creating a personal turbine which will be able to meet all the demands of a modern person's electrical needs, just as a large turbine can meet the electricity demands of a small city.

Problems have occurred with heat dissipation and high-speed bearings in these new microturbines. Moreover, their expected efficiency is a very low 5-6%. According to Professor Epstein, current commercial Li-ion rechargeable batteries deliver about 120-150 W·h/kg. MIT's millimeter size turbine will deliver 500-700 W·h/kg in the near term, rising to 1200-1500 W·h/kg in the longer term.

A similar microturbine built in Belgium has a rotor diameter of 20 mm and is expected to produce about 1000 W.

External Combustion

Most gas turbines are internal combustion engines but it is also possible to manufacture an external combustion gas turbine which is, effectively, a turbine version of a hot air engine. Those systems are usually indicated as EFGT (Externally Fired Gas Turbine) or IFGT (Indirectly Fired Gas Turbine).

External combustion has been used for the purpose of using pulverized coal or finely ground biomass (such as sawdust) as a fuel. In the indirect system, a heat exchanger is used and only clean air with no combustion products travels through the power turbine. The thermal efficiency is lower in the indirect type of external combustion; however, the turbine blades are not subjected to combustion products and much lower quality (and therefore cheaper) fuels are able to be used.

When external combustion is used, it is possible to use exhaust air from the turbine as the primary combustion air. This effectively reduces global heat losses, although heat losses associated with the combustion exhaust remain inevitable.

Closed-cycle gas turbines based on helium or supercritical carbon dioxide also hold promise for use with future high temperature solar and nuclear power generation.

In Surface Vehicles

Gas turbines are often used on ships, locomotives, helicopters, tanks, and to a lesser extent, on cars, buses, and motorcycles.

The 1967 *STP Oil Treatment Special* on display at the Indianapolis Motor Speedway Hall of Fame Museum, with the Pratt & Whitney gas turbine shown

A 1968 Howmet TX, the only turbine-powered race car to have won a race

A key advantage of jets and turboprops for aeroplane propulsion - their superior performance at high altitude compared to piston engines, particularly naturally aspirated ones - is irrelevant in most automobile applications. Their power-to-weight advantage, though less critical than for aircraft, is still important.

Gas turbines offer a high-powered engine in a very small and light package. However, they are not as responsive and efficient as small piston engines over the wide range of RPMs and powers needed in vehicle applications. In series hybrid vehicles, as the driving electric motors are mechanically detached from the electricity generating engine, the responsiveness, poor performance at low speed and low efficiency at low output problems are much less important. The turbine can be run at optimum speed for its power output, and batteries and ultracapacitors can supply power as needed, with the engine cycled on and off to run it only at high efficiency. The emergence of the continuously variable transmission may also alleviate the responsiveness problem.

Turbines have historically been more expensive to produce than piston engines, though this is partly because piston engines have been mass-produced in huge quantities for decades, while small gas turbine engines are rarities; however, turbines are mass-produced in the closely related form of the turbocharger.

The turbocharger is basically a compact and simple free shaft radial gas turbine which is driven by the piston engine's exhaust gas. The centripetal turbine wheel drives a centrifugal compressor wheel through a common rotating shaft. This wheel supercharges the engine air intake to a degree that can be controlled by means of a wastegate or by dynamically modifying the turbine housing's geometry (as in a VGT turbocharger). It mainly serves as a power recovery device which converts a great deal of otherwise wasted thermal and kinetic energy into engine boost.

Turbo-compound engines (actually employed on some trucks) are fitted with blow down turbines which are similar in design and appearance to a turbocharger except for the turbine shaft being mechanically or hydraulically connected to the engine's crankshaft instead of to a centrifugal compressor, thus providing additional power instead of boost. While the turbocharger is a pressure turbine, a power recovery turbine is a velocity one.

Passenger Road Vehicles (Cars, Bikes, and Buses)

A number of experiments have been conducted with gas turbine powered automobiles, the largest by Chrysler. More recently, there has been some interest in the use of turbine engines for hybrid electric cars. For instance, a consortium led by micro gas turbine company Bladon Jets has secured investment from the Technology Strategy Board to develop an Ultra Lightweight Range Extender (ULRE) for next generation electric vehicles. The objective of the consortium, which includes luxury car maker Jaguar Land Rover and leading electrical machine company SR Drives, is to produce the world's first commercially viable - and environmentally friendly - gas turbine generator designed specifically for automotive applications.

The common turbocharger for gasoline or diesel engines is also a turbine derivative.

Concept Cars

The first serious investigation of using a gas turbine in cars was in 1946 when two engineers, Robert Kafka and Robert Engerstein of Carney Associates, a New York engineering firm, came up with the concept where a unique compact turbine engine design would provide power for a rear wheel drive car. After an article appeared in *Popular Science*, there was no further work, beyond the paper stage.

The 1950 Rover JET1

In 1950, designer F.R. Bell and Chief Engineer Maurice Wilks from British car manufacturers Rover unveiled the first car powered with a gas turbine engine. The two-seater JET1 had the engine positioned behind the seats, air intake grilles on either side of the car, and exhaust outlets on the top of the tail. During tests, the car reached top speeds of 140 km/h (87 mph), at a turbine speed of 50,000 rpm. The car ran on petrol, paraffin (kerosene) or diesel oil, but fuel consumption problems proved insurmountable for a production car. It is on display at the London Science Museum.

A French turbine powered car, the Socema-Gregoire, was displayed at the October 1952 Paris Auto Show. It was designed by the French engineer Jean-Albert Grégoire.

Firebird I

The first turbine powered car built in the US was the GM Firebird I which began evaluations in 1953. While photos of the Firebird I may suggest that the jet turbine's thrust propelled the car like an aircraft, the turbine in fact drove the rear wheels. The Firebird 1 was never meant as a serious commercial passenger car and was solely built for testing & evaluation as well as public relation purposes.

Engine compartment of a Chrysler 1963 Turbine car

Starting in 1954 with a modified Plymouth, the American car manufacturer Chrysler demonstrated several prototype gas turbine-powered cars from the early 1950s through the early 1980s. Chrysler built fifty Chrysler Turbine Cars in 1963 and conducted the only consumer trial of gas turbine-powered cars. Each of their turbines employed a unique rotating recuperator, referred to as a regenerator that increased efficiency.

In 1954 FIAT unveiled a concept car with a turbine engine, called Fiat Turbina. This vehicle, looking like an aircraft with wheels, used a unique combination of both jet thrust and the engine driving the wheels. Speeds of 282 km/h (175 mph) were claimed.

The original General Motors Firebird was a series of concept cars developed for the 1953, 1956 and 1959 Motorama auto shows, powered by gas turbines.

As a result of the U.S. Clean Air Act Amendments of 1970, research was funded to developing automotive gas turbine technology. Design concepts and vehicles were conducted by Chrysler, General Motors, Ford (in collaboration with AiResearch), and American Motors (in conjunction with Williams Research). Long-term tests were conducted evaluate comparable cost efficiency. Several AMC Hornets were powered by a small Williams regenerative gas turbines weighing 250 lb (113 kg) and producing 80 hp (60 kW; 81 PS) at 4450 rpm.

Toyota demonstrated several gas turbine powered concept cars, such as the Century gas turbine hybrid in 1975, the Sports 800 Gas Turbine Hybrid in 1979 and the GTV in 1985. No production vehicles were made. The GT24 engine was exhibited in 1977 without a vehicle.

In the early 1990s Volvo introduced the Volvo Environmental Concept Car(ECC) which was a gas turbine powered hybrid car.

In 1993 General Motors introduced the first commercial gas turbine powered hybrid vehicle—as a limited production run of the EV-1 series hybrid. A Williams International 40 kW turbine drove an alternator which powered the battery-electric powertrain. The turbine design included a recuperator. Later on in 2006 GM went into the EcoJet concept car project with Jay Leno.

At the 2010 Paris Motor Show Jaguar demonstrated its Jaguar C-X75 concept car. This electrically powered supercar has a top speed of 204 mph (328 km/h) and can go from 0 to 62 mph (0 to 100 km/h) in 3.4 seconds. It uses Lithium-ion batteries to power 4 electric motors which combine to produce some 780 bhp. It will do 68 miles (109 km) on a single charge of the batteries, but in addition it uses a pair of Bladon Micro Gas Turbines to re-charge the batteries extending the range to 560 miles (900 km).

Racing Cars

The first race car (in concept only) fitted with a turbine was in 1955 by a US Air Force group as a hobby project with a turbine loaned them by Boeing and a race car owned by Firestone Tire & Rubber company. The first race car fitted with a turbine for the goal of actual racing was by Rover and the BRM Formula One team joined forces to produce the Rover-BRM, a gas turbine powered coupe, which entered the 1963 24 Hours of Le Mans, driven by Graham Hill and Richie Ginther. It averaged 107.8 mph (173.5 km/h) and had a top speed of 142 mph (229 km/h). American Ray Heppenstall joined Howmet Corporation and McKee Engineering together to develop their own gas turbine sports car in 1968, the Howmet TX, which ran several American and European events, including two wins, and also participated in the 1968 24 Hours of Le Mans. The cars used Continental gas turbines, which eventually set six FIA land speed records for turbine-powered cars.

For open wheel racing, 1967's revolutionary STP-Paxton Turbocar fielded by racing and entrepreneurial legend Andy Granatelli and driven by Parnelli Jones nearly won the Indianapolis 500; the Pratt & Whitney ST6B-62 powered turbine car was almost a lap ahead of the second place car when a gearbox bearing failed just three laps from the finish line. The next year the STP Lotus 56 turbine car won the Indianapolis 500 pole position even though new rules restricted the air intake dramatically. In 1971 Lotus principal Colin Chapman introduced the Lotus 56B F1 car, powered by a Pratt & Whitney STN 6/76 gas turbine. Chapman had a reputation of building radical championship-winning cars, but had to abandon the project because there were too many problems with turbo lag.

Buses

The arrival of the Capstone Microturbine has led to several hybrid bus designs, starting with HEV-1 by AVS of Chattanooga, Tennessee in 1999, and closely followed by Ebus and ISE Research in California, and DesignLine Corporation in New Zealand (and later the United States). AVS turbine hybrids were plagued with reliability and quality control problems, resulting in liquidation of AVS in 2003. The most successful design by Designline is now operated in 5 cities in 6 countries, with over 30 buses in operation worldwide, and order for several hundred being delivered to Baltimore, and NYC.

Brescia Italy is using serial hybrid buses powered by microturbines on routes through the historical sections of the city.

Motorcycles

The MTT Turbine Superbike appeared in 2000 (hence the designation of Y2K Superbike by MTT) and is the first production motorcycle powered by a turbine engine - specifically, a Rolls-Royce Allison model 250 turboshaft engine, producing about 283 kW (380 bhp). Speed-tested to 365 km/h or 227 mph (according to some stories, the testing team ran out of road during the test), it holds the Guinness World Record for most powerful production motorcycle and most expensive production motorcycle, with a price tag of US$185,000.

Trains

Several locomotive classes have been powered by gas turbines, the most recent incarnation being Bombardier's JetTrain.

Tanks

Marines from 1st Tank Battalion load a Honeywell AGT1500 multi-fuel turbine back into an M1 Abrams tank at Camp Coyote, Kuwait, February 2003

The German Army's development division, the Heereswaffenamt (Army Ordnance Board), studied a number of gas turbine engines for use in tanks starting in mid-1944. The first gas turbine engines used for armoured fighting vehicle GT 101 was installed in the Panther tank. The second use of a gas turbine in an armoured fighting vehicle was in 1954 when a unit, PU2979, specifically developed for tanks by C. A. Parsons & Co., was installed and trialled in a British Conqueror tank. The Stridsvagn 103 was developed in the 1950s and was the first mass-produced main battle tank to use a turbine engine. Since then, gas turbine engines have been used as APUs in some tanks and as main powerplants in Soviet/Russian T-80s and U.S. M1 Abrams tanks, among others. They are lighter and smaller than diesels at the same sustained power output but the models installed to date are less fuel efficient than the equivalent diesel, especially at idle, requiring more fuel to achieve the same combat range. Successive models of M1 have addressed this problem with battery packs or secondary generators to power the tank's systems while stationary, saving fuel by reducing the need to idle the main turbine. T-80s can mount three large external fuel drums to extend their range. Russia has stopped production of the T-80 in favour of the diesel-powered T-90 (based on the T-72), while Ukraine has developed the diesel-powered T-80UD and T-84 with nearly the power of the gas-turbine tank. The French Leclerc MBT's diesel powerplant features the

"Hyperbar" hybrid supercharging system, where the engine's turbocharger is completely replaced with a small gas turbine which also works as an assisted diesel exhaust turbocharger, enabling engine RPM-independent boost level control and a higher peak boost pressure to be reached (than with ordinary turbochargers). This system allows a smaller displacement and lighter engine to be used as the tank's powerplant and effectively removes turbo lag. This special gas turbine/turbocharger can also work independently from the main engine as an ordinary APU.

A turbine is theoretically more reliable and easier to maintain than a piston engine, since it has a simpler construction with fewer moving parts but in practice turbine parts experience a higher wear rate due to their higher working speeds. The turbine blades are highly sensitive to dust and fine sand, so that in desert operations air filters have to be fitted and changed several times daily. An improperly fitted filter, or a bullet or shell fragment that punctures the filter, can damage the engine. Piston engines (especially if turbocharged) also need well-maintained filters, but they are more resilient if the filter does fail.

Like most modern diesel engines used in tanks, gas turbines are usually multi-fuel engines.

Marine Applications
Naval

The Gas turbine from MGB 2009

Gas turbines are used in many naval vessels, where they are valued for their high power-to-weight ratio and their ships' resulting acceleration and ability to get underway quickly.

The first gas-turbine-powered naval vessel was the Royal Navy's Motor Gun Boat *MGB 2009* (formerly *MGB 509*) converted in 1947. Metropolitan-Vickers fitted their F2/3 jet engine with a power turbine. The Steam Gun Boat *Grey Goose* was converted to Rolls-Royce gas turbines in 1952 and operated as such from 1953. The Bold class Fast Patrol Boats *Bold Pioneer* and *Bold Pathfinder* built in 1953 were the first ships created specifically for gas turbine propulsion.

The first large scale, partially gas-turbine powered ships were the Royal Navy's Type 81 (Tribal class) frigates with combined steam and gas powerplants. The first, HMS *Ashanti* was commissioned in 1961.

The German Navy launched the first *Köln*-class frigate in 1961 with 2 Brown, Boveri & Cie gas turbines in the world's first combined diesel and gas propulsion system.

The Danish Navy had 6 *Søløven*-class torpedo boats (the export version of the British Brave class fast patrol boat) in service from 1965 to 1990, which had 3 Bristol Proteus (later RR Proteus) Marine Gas Turbines rated at 9,510 kW (12,750 shp) combined, plus two General Motors Diesel engines, rated at 340 kW (460 shp), for better fuel economy at slower speeds. And they also produced 10 Willemoes Class Torpedo / Guided Missile boats (in service from 1974 to 2000) which had 3 Rolls Royce Marine Proteus Gas Turbines also rated at 9,510 kW (12,750 shp), same as the Søløven-class boats, and 2 General Motors Diesel Engines, rated at 600 kW (800 shp), also for improved fuel economy at slow speeds.

The Swedish Navy produced 6 Spica-class torpedo boats between 1966 and 1967 powered by 3 Bristol Siddeley Proteus 1282 turbines, each delivering 3,210 kW (4,300 shp). They were later joined by 12 upgraded Norrköping class ships, still with the same engines. With their aft torpedo tubes replaced by antishipping missiles they served as missile boats until the last was retired in 2005.

The Finnish Navy commissioned two *Turunmaa*-class corvettes, *Turunmaa* and *Karjala*, in 1968. They were equipped with one 16,410 kW (22,000 shp) Rolls-Royce Olympus TM1 gas turbine and three Wärtsilä marine diesels for slower speeds. They were the fastest vessels in the Finnish Navy; they regularly achieved speeds of 35 knots, and 37.3 knots during sea trials. The *Turunmaa*s were paid off in 2002. *Karjala* is today a museum ship in Turku, and *Turunmaa* serves as a floating machine shop and training ship for Satakunta Polytechnical College.

The next series of major naval vessels were the four Canadian *Iroquois*-class helicopter carrying destroyers first commissioned in 1972. They used 2 ft-4 main propulsion engines, 2 ft-12 cruise engines and 3 Solar Saturn 750 kW generators.

An LM2500 gas turbine on USS *Ford*

The first U.S. gas-turbine powered ship was the U.S. Coast Guard's *Point Thatcher*, a cutter commissioned in 1961 that was powered by two 750 kW (1,000 shp) turbines utilizing controllable-pitch propellers. The larger *Hamilton*-class High Endurance Cutters, was the first class of larger cutters to utilize gas turbines, the first of which (USCGC *Hamilton*) was commissioned in 1967. Since then, they have powered the U.S. Navy's *Oliver Hazard Perry*-class frigates, *Spruance* and *Arleigh Burke*-class destroyers, and *Ticonderoga*-class guided missile cruisers. USS *Makin Island*, a modified *Wasp*-class amphibious assault ship, is to be the Navy's first amphibious assault ship powered by gas turbines. The marine gas turbine operates in a more corrosive atmosphere due to presence of sea salt in air and fuel and use of cheaper fuels.

Civilian Maritime

Up to the late 1940s much of the progress on marine gas turbines all over the world took place in design offices and engine builder's workshops and development work was led by the British Royal Navy and other Navies. While interest in the gas turbine for marine purposes, both naval and mercantile, continued to increase, the lack of availability of the results of operating experience on early gas turbine projects limited the number of new ventures on seagoing commercial vessels being embarked upon. In 1951, the Diesel-electric oil tanker *Auris*, 12,290 Deadweight tonnage (DWT) was used to obtain operating experience with a main propulsion gas turbine under service conditions at sea and so became the first ocean-going merchant ship to be powered by a gas turbine. Built by Hawthorn Leslie at Hebburn-on-Tyne, UK, in accordance with plans and specifications drawn up by the Anglo-Saxon Petroleum Company and launched on the UK's Princess Elizabeth's 21st birthday in 1947, the ship was designed with an engine room layout that would allow for the experimental use of heavy fuel in one of its high-speed engines, as well as the future substitution of one of its diesel engines by a gas turbine. The *Auris* operated commercially as a tanker for three-and-a-half years with a diesel-electric propulsion unit as originally commissioned, but in 1951 one of its four 824 kW (1,105 bhp) diesel engines – which were known as "Faith", "Hope", "Charity" and "Prudence" - was replaced by the world's first marine gas turbine engine, a 890 kW (1,200 bhp) open-cycle gas turbo-alternator built by British Thomson-Houston Company in Rugby. Following successful sea trials off the Northumbrian coast, the *Auris* set sail from Hebburn-on-Tyne in October 1951 bound for Port Arthur in the US and then Curacao in the southern Caribbean returning to Avonmouth after 44 days at sea, successfully completing her historic trans-Atlantic crossing. During this time at sea the gas turbine burnt diesel fuel and operated without an involuntary stop or mechanical difficulty of any kind. She subsequently visited Swansea, Hull, Rotterdam, Oslo and Southampton covering a total of 13,211 nautical miles. The *Auris* then had all of its power plants replaced with a 3,910 kW (5,250 shp) directly coupled gas turbine to become the first civilian ship to operate solely on gas turbine power.

Despite the success of this early experimental voyage the gas turbine was not to replace the diesel engine as the propulsion plant for large merchant ships. At constant cruising speeds the diesel engine simply had no peer in the vital area of fuel economy. The gas turbine did have more success in Royal Navy ships and the other naval fleets of the world where sudden and rapid changes of speed are required by warships in action.

The United States Maritime Commission were looking for options to update WWII Liberty ships, and heavy-duty gas turbines were one of those selected. In 1956 the *John Sergeant* was lengthened and equipped with a General Electric 4,900 kW (6,600 shp) HD gas turbine with exhaust-gas regeneration, reduction gearing and a variable-pitch propeller. It operated for 9,700 hours using residual fuel(Bunker C) for 7,000 hours. Fuel efficiency was on a par with steam propulsion at 0.318 kg/kW (0.523 lb/hp) per hour, and power output was higher than expected at 5,603 kW (7,514 shp) due to the ambient temperature of the North Sea route being lower than the design temperature of the gas turbine. This gave the ship a speed capability of 18 knots, up from 11 knots with the original power plant, and well in excess of the 15 knot targeted. The ship made its first transatlantic crossing with an average speed of 16.8 knots, in spite of some rough weather along the way. Suitable Bunker C fuel was only available at limited ports because the quality of the fuel was of a critical nature. The fuel oil also had to be treated on board to reduce contaminants and this was a labor-intensive process that was not suitable for automation at the time. Ultimately, the variable-pitch propeller, which was of a new and untested design, ended the trial, as three consecutive annual inspections revealed stress-cracking. This did not reflect poorly on the marine-propulsion gas-turbine concept though, and the trial was a success overall. The success of this trial opened the way for more development by GE on the use of HD gas turbines for marine use with heavy fuels. The *John Sergeant* was scrapped in 1972 at Portsmouth PA.

Boeing Jetfoil 929-100-007 *Urzela* of TurboJET

Boeing launched its first passenger-carrying waterjet-propelled hydrofoil Boeing 929, in April 1974. Those ships were powered by twin Allison gas turbines of the KF-501 series.

Between 1971 and 1981, Seatrain Lines operated a scheduled container service between ports on the eastern seaboard of the United States and ports in northwest Europe across the North Atlantic with four container ships of 26,000 tonnes DWT. Those ships were powered by twin Pratt & Whitney gas turbines of the FT 4 series. The four ships in the class were named *Euroliner*, *Eurofreighter*, *Asialiner* and *Asiafreighter*. Following the dramatic Organization of the Petroleum Exporting Countries (OPEC) price increases of the mid-1970s, operations were constrained by rising fuel costs. Some modification of the engine systems on those ships was undertaken to permit the burning of a lower grade of fuel (i.e., marine diesel). Reduction of fuel costs was successful using a different untested fuel in a marine gas turbine but maintenance costs increased with the fuel change. After 1981 the ships were sold and refitted with, what at the time, was more economical diesel-fueled engines but the increased engine size reduced cargo space.

The first passenger ferry to use a gas turbine was the GTS *Finnjet*, built in 1977 and powered by two Pratt & Whitney FT 4C-1 DLF turbines, generating 55,000 kW (74,000 shp) and propelling the ship to a speed of 31 knots. However, the Finnjet also illustrated the shortcomings of gas turbine propulsion in commercial craft, as high fuel prices made operating her unprofitable. After four years of service additional diesel engines were installed on the ship to reduce running costs during the off-season. The Finnjet was also the first ship with a Combined diesel-electric and gas propulsion. Another example of commercial usage of gas turbines in a passenger ship is Stena Line's HSS class fastcraft ferries. HSS 1500-class *Stena Explorer*, *Stena Voyager* and *Stena Discovery* vessels use combined gas and gas setups of twin GE LM2500 plus GE LM1600 power for a total of 68,000 kW (91,000 shp). The slightly smaller HSS 900-class *Stena Carisma*, uses twin

ABB–STAL (sv) GT35 turbines rated at 34,000 kW (46,000 shp) gross. The *Stena Discovery* was withdrawn from service in 2007, another victim of too high fuel costs.

In July 2000 the *Millennium* became the first cruise ship to be propelled by gas turbines, in a Combined Gas and Steam Turbine configuration. The liner RMS Queen Mary 2 uses a Combined Diesel and Gas Turbine configuration.

In marine racing applications the 2010 C5000 Mystic catamaran Miss GEICO uses two Lycoming T-55 turbines for its power system.

Advances in Technology

Gas turbine technology has steadily advanced since its inception and continues to evolve. Development is actively producing both smaller gas turbines and more powerful and efficient engines. Aiding in these advances are computer based design (specifically CFD and finite element analysis) and the development of advanced materials: Base materials with superior high temperature strength (e.g., single-crystal superalloys that exhibit yield strength anomaly) or thermal barrier coatings that protect the structural material from ever higher temperatures. These advances allow higher compression ratios and turbine inlet temperatures, more efficient combustion and better cooling of engine parts.

Computational Fluid Dynamics (CFD) has contributed to substantial improvements in the performance and efficiency of Gas Turbine engine components through enhanced understanding of the complex viscous flow and heat transfer phenomena involved. For this reason, CFD is one of the key computational tool used in Design & development of gas turbine engines.

The simple-cycle efficiencies of early gas turbines were practically doubled by incorporating inter-cooling, regeneration (or recuperation), and reheating. These improvements, of course, come at the expense of increased initial and operation costs, and they cannot be justified unless the decrease in fuel costs offsets the increase in other costs. The relatively low fuel prices, the general desire in the industry to minimize installation costs, and the tremendous increase in the simple-cycle efficiency to about 40 percent left little desire for opting for these modifications.

On the emissions side, the challenge is to increase turbine inlet temperatures while at the same time reducing peak flame temperature in order to achieve lower NOx emissions and meet the latest emission regulations. In May 2011, Mitsubishi Heavy Industries achieved a turbine inlet temperature of 1,600 °C on a 320 megawatt gas turbine, and 460 MW in gas turbine combined-cycle power generation applications in which gross thermal efficiency exceeds 60%.

Compliant foil bearings were commercially introduced to gas turbines in the 1990s. These can withstand over a hundred thousand start/stop cycles and have eliminated the need for an oil system. The application of microelectronics and power switching technology have enabled the development of commercially viable electricity generation by micro turbines for distribution and vehicle propulsion.

Advantages and Disadvantages

The following are advantages and disadvantages of gas-turbine engines:

Advantages

- Very high power-to-weight ratio, compared to reciprocating engines

- Smaller than most reciprocating engines of the same power rating

- Moves in one direction only, with far less vibration than a reciprocating engine

- Fewer moving parts than reciprocating engines implies lower maintenance cost.

- Greater reliability, particularly in applications where sustained high power output is required

- Waste heat is dissipated almost entirely in the exhaust. This results in a high temperature exhaust stream that is very usable for boiling water in a combined cycle, or for cogeneration

- Low operating pressures

- High operation speeds

- Low lubricating oil cost and consumption

- Can run on a wide variety of fuels

- Very low toxic emissions of CO and HC due to excess air, complete combustion and no "quench" of the flame on cold surfaces

Disadvantages

- Cost is very high

- Less efficient than reciprocating engines at idle speed

- Longer startup than reciprocating engines

- Less responsive to changes in power demand compared with reciprocating engines

- Characteristic whine can be hard to suppress

Testing

British, German, other national and international test codes are used to standardize the procedures and definitions used to test gas turbines. Selection of the test code to be used is an agreement between the purchaser and the manufacturer, and has some significance to the design of the turbine and associated systems. In the United States, ASME has produced several performance test codes on gas turbines. This includes ASME PTC 22-2014. These ASME performance test codes have gained international recognition and acceptance for testing gas turbines. The single most important and differentiating characteristic of ASME performance test codes, including PTC 22, is that the test uncertainty of the measurement indicates the quality of the test and is not to be used as a commercial tolerance.

Wankel Engine

A cut-away of a Wankel engine shown at the Deutsches Museum in Munich, Germany

The Mazda RX-8, a sports car powered by a Wankel engine

Norton Classic air-cooled twin-rotor motorcycle

The Wankel engine is a type of internal combustion engine using an eccentric rotary design to convert pressure into rotating motion. In contrast to the more common reciprocating piston designs, the Wankel engine delivers advantages of simplicity, smoothness, compactness, high revolutions per minute, and a high power-to-weight ratio. The engine is commonly referred to as a rotary engine, although this name applies also to other completely different designs. All parts rotate moving in one direction, as opposed to the common reciprocating piston engine which has pistons violently changing direction. The four-stroke cycle occurs in a moving combustion chamber between the inside of an oval-like epitrochoid-shaped housing, and a rotor that is similar in shape to a Reuleaux triangle with sides that are somewhat flatter.

The concept of the engine was conceived by German engineer Felix Wankel. Wankel received his first patent for the engine in 1929, began development in the early 1950s at NSU, and completed a working prototype in 1957. NSU subsequently licensed the design to companies around the world, who have continually added improvements.

The Wankel engine has the advantages of compact design and low weight over the most commonly used internal combustion engine employing reciprocating pistons. These advantages have given rotary engine applications in a variety of vehicles and devices, including: automobiles, motorcycles, racing cars, aircraft, go-karts, jet skis, snowmobiles, chain saws, and auxiliary power units. The point of power to weight has been reached of under one pound weight per horsepower output.

History

Early Developments

The first DKM Wankel engine designed by Felix Wankel, the DKM 54 (*Drehkolbenmotor*), at the Deutsches Museum in Bonn, Germany: the rotor and its housing spin.

The first KKM Wankel Engine designed by Hanns Dieter Paschke, the NSU KKM 57P (*Kreiskolbenmotor*), at Autovision und Forum, Germany: the rotor housing is static.

In 1951, NSU Motorenwerke AG in Germany began development of the engine with two models being developed. The first, the DKM motor, was developed by the engineer Felix Wankel. The second, the KKM motor, was developed by Hanns Dieter Paschke, which was adopted forming the modern Wankel engine. The Wankel engine design used today was not designed by Felix Wankel. Titling the engine "the Paschke engine" could be considered to be more apt.

The basis of the DKM type of motor is that both the rotor and the housing spin around on separate axes. The DKM motor reached higher revolutions per minute and was more naturally balanced. However, the engine needed to be stripped to change the spark plugs and contained more parts. The KKM engine is simpler, having a fixed housing.

The first working prototype, DKM 54, produced 21 horsepower and ran on February 1, 1957, at the NSU research and development department *Versuchsabteilung TX*. The KKM 57 (the Wankel rotary engine, *Kreiskolbenmotor*) was constructed by NSU engineer Hanns Dieter Paschke in 1957 without the knowledge of Felix Wankel, who later remarked "you have turned my race horse into a plow mare".

Licenses Issued

In 1960, NSU, the firm that employed the two inventors, and the US firm Curtiss-Wright, signed a joint agreement. NSU were to concentrate on low- and medium-powered Wankel engine development and Curtiss-Wright developing high-powered engines, including aircraft engines of which Curtiss-Wright had decades of experience designing and producing. Curtiss-Wright recruited Max Bentele to head their design team.

Many manufacturers signed license agreements for development, attracted by the smoothness, quiet running, and reliability emanating from the uncomplicated design. Amongst them were Alfa Romeo, American Motors, Citroen, Ford, General Motors, Mazda, Mercedes-Benz, Nissan, Porsche, Rolls-Royce, Suzuki, and Toyota. In the United States in 1959, under license from NSU, Curtiss-Wright pioneered improvements in the basic engine design. In Britain, in the 1960s, Rolls Royce's Motor Car Division pioneered a two-stage diesel version of the Wankel engine.

Citroën did much research, producing the M35 and GS Birotor, using engines produced by Comotor, a joint venture of Citroën and NSU. General Motors seemed to have concluded the Wankel engine was slightly more expensive to build than an equivalent reciprocating engine. General Motors claimed to have solved the fuel economy issue, but failed in obtaining in a concomitant way acceptable exhaust emissions. Mercedes-Benz fitted a Wankel engine in their C111 concept car.

Deere & Company designed a version that was capable of using a variety of fuels. The design was proposed as the power source for United States Marine Corps combat vehicles and other equipment in the late 1980s.

In 1961, the Soviet research organization of NATI, NAMI, and VNIImotoprom commenced development creating experimental engines with different technologies. Soviet automobile manufacturer AvtoVAZ also experimented in Wankel engine design without a license, introducing a limited number of engines in some cars.

Despite much research and development throughout the world, only Mazda has produced Wankel engines in large quantities.

Developments for Motorcycles

In Britain, Norton Motorcycles developed a Wankel rotary engine for motorcycles, based on the Sachs air-cooled rotor Wankel that powered the DKW/Hercules W-2000 motorcycle, this two-rotor engine was included in their Commander and F1. Norton improved on the Sachs's air cooling, introducing a plenum chamber. Suzuki also made a production motorcycle powered by a Wankel engine, the RE-5, using ferroTiC alloy apex seals and an NSU rotor in a successful attempt to prolong the engine's life.

Developments for Cars

Mazda and NSU signed a study contract to develop the Wankel engine in 1961 and competed to bring the first Wankel-powered automobile to market. Although Mazda produced an experimental Wankel that year, NSU was first with a Wankel automobile for sale, the sporty NSU Spider in 1964; Mazda countered with a display of two- and four-rotor Wankel engines at that year's Tokyo Motor Show. In 1967, NSU began production of a Wankel-engined luxury car, the Ro 80. However, NSU had not produced reliable apex seals on the rotor, unlike Mazda and Curtiss-Wright. NSU had problems with apex seals' wear, poor shaft lubrication, and poor fuel economy, leading to frequent engine failures, not solved until 1972, which led to large warranty costs curtailing further NSU Wankel engine development. This premature release of the new Wankel engine gave a poor reputation for all makes and even when these issues were solved in the last engines produced by NSU in the second half of the '70s, sales did not recover. Audi, after the takeover of NSU, built in 1979 a new KKM 871 engine with side intake ports and 750 cc per chamber, 170 HP at 6,500 rpm, and 220 Nm at 3,500 rpm. The engine was installed in an Audi 100 hull they named "Audi 200", but the engine was not mass-produced.

Mazda's first Wankel engine, at the Mazda Museum in Hiroshima, Japan

Mazda, however, claimed to have solved the apex seal problem, and operated test engines at high speed for 300 hours without failure. After years of development, Mazda's first Wankel engine car was the 1967 Cosmo 110S. The company followed with a number of Wankel ("rotary" in the company's terminology) vehicles, including a bus and a pickup truck. Customers often cited the cars' smoothness of operation. However, Mazda chose a method to comply with hydrocarbon emission standards that, while less expensive to produce, increased fuel consumption. Unfortunately for Mazda, this was introduced immediately prior to a sharp rise in fuel prices. Curtiss-Wright produced the RC2-60 engine which was comparable to a V8 engine in performance and fuel consumption. Unlike NSU, by 1966 Curtiss-Wright had solved the rotor sealing issue with seals lasting 100,000 miles.

Mazda later abandoned the Wankel in most of their automotive designs, continuing to use the engine in their sports car range only, producing the RX-7 until August 2002. The company normally used two-rotor designs. A more advanced twin-turbo three-rotor engine was fitted in the 1991 Eunos Cosmo sports car. In 2003, Mazda introduced the Renesis engine fitted in the RX-8. The Renesis engine relocated the ports for exhaust from the periphery of the rotary housing to the sides, allowing for larger overall ports, better airflow, and further power gains. Some early Wankel engines had also side exhaust ports, the concept being abandoned because of carbon buildup in ports and the sides of the rotor. The

Renesis engine solved the problem by using a keystone scraper side seal, and approached the thermal distortion difficulties by adding some parts made of ceramics. The Renesis is capable of 238 hp (177 kW) with improved fuel economy, reliability, and lower emissions than previous Mazda rotary engines, all from a nominal 1.3 L displacement. However, this was not enough to meet more stringent emissions standards. Mazda ended production of their Wankel engine in 2012 after the engine failed to meet the improved Euro 5 emission standards, leaving no automotive company selling a Wankel-powered vehicle. The company is continuing development of the next generation of Wankel engines, the SkyActiv-R with a new rear wheel drive sports car model announced in October 2015 although with no launch date given. Mazda states that the SkyActiv-R solves the three key issues with previous rotary engines: fuel economy, emissions and reliability. Mazda announced the introduction of the series-hybrid Mazda2 EV car using a Wankel engine as a range extender.

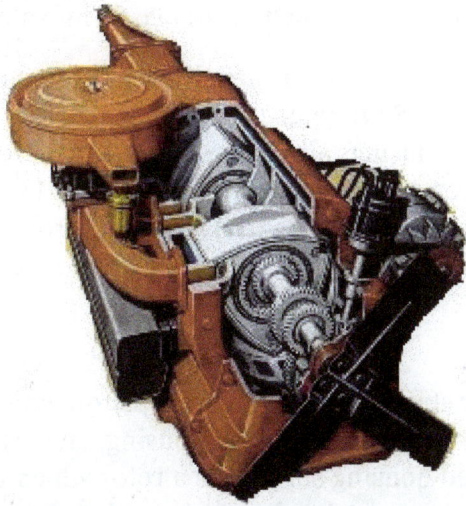

1972 GM Rotary engine cutaway shows twin-rotors

American Motors (AMC) was so convinced "... that the rotary engine will play an important role as a powerplant for cars and trucks of the future....", that the chairman, Roy D. Chapin Jr., of the smallest U.S. automaker signed an agreement in February 1973, after a year's negotiations, to build Wankels for both passenger cars and Jeeps, as well as the right to sell any rotary engines it produced to other companies. American Motors' president, William Luneburg, did not expect dramatic development through to 1980, however Gerald C. Meyers, AMC's vice president of the engineering product group, suggested that AMC should buy the engines from Curtiss-Wright before developing its own Wankel engines and predicted a total transition to rotary power by 1984. Plans called for the engine to be used in the AMC Pacer, but development was pushed back. American Motors designed the unique Pacer around the engine. By 1974, AMC had decided to purchase the General Motors Wankel instead of building an engine in-house. Both General Motors and AMC confirmed the relationship would benefit in marketing the new engine, with AMC claiming that the General Motors' Wankel achieved good fuel economy. However, General Motors' engines had not reached production when the Pacer was launched onto the market. The 1973 oil crisis played a part in frustrating the uptake of the Wankel engine. Rising fuel prices and talk about proposed US emission standards legislation also added to the concerns.

By 1974, General Motors R&D had not succeeded in producing a Wankel engine meeting both the emission requirements and good fuel economy, leading the company to consider cancelling the project. As General Motors managers were cancelling the Wankel project, the R&D team released only partly the results of their most recent research, which claimed to have solved the fuel economy problem, and building reliable engines with a lifespan above 530,000 miles. These findings were not taken into account when the cancellation order was issued. The cancellation of General Motors' Wankel project required AMC to reconfigure the Pacer to house its venerable AMC straight-6 engine driving the rear-wheels.

In 1974, the Soviets created a special engine design bureau, which, in 1978, designed an engine designated as "VAZ-311". In 1980, the company commenced delivery of the VAZ-411 twin-rotor Wankel engine in VAZ-2106s and Lada cars, with about 200 manufactured. Most of the production went to the security services. The next models were the VAZ-4132 and VAZ-415. Aviadvigatel, the Soviet aircraft engine design bureau, is known to have produced Wankel engines with electronic injection for aircraft and helicopters, though little specific information has surfaced.

Ford conducted research in Wankel engines, resulting in patents granted: GB 1460229, 1974, method for fabricating housings; US 3833321 1974, side plates coating; US 3890069, 1975, housing coating; CA 1030743, 1978: Housings alignment; CA 1045553, 1979, Reed-Valve assembly. Mr. Henry Ford II 1972 statement regarding the production of a Ford Wankel engine was: 'The Rotary probably won't replace the piston in my lifetime". (Harris Edward Dark, 'the Wankel rotary engine, introduction and guide'. Indiana Uiversity Press 1974, pag 80, ISBN 0-253-19021-5).

Design

In the Wankel engine, the four strokes of an Otto cycle piston engine occur in the space between a three-sided symmetric rotor and the inside of a housing. In each rotor of the Wankel engine, the oval-like epitrochoid-shaped housing surrounds a rotor which is triangular with bow-shaped flanks (often confused with a Reuleaux triangle, a three-pointed curve of constant width, but with the bulge in the middle of each side a bit more flattened). The theoretical shape of the rotor between the fixed corners is the result of a minimization of the volume of the geometric combustion chamber and a maximization of the compression ratio, respectively. The symmetric curve connecting two arbitrary apexes of the rotor is maximized in the direction of the inner housing shape with the constraint that it not touch the housing at any angle of rotation (an arc is not a solution of this optimization problem).

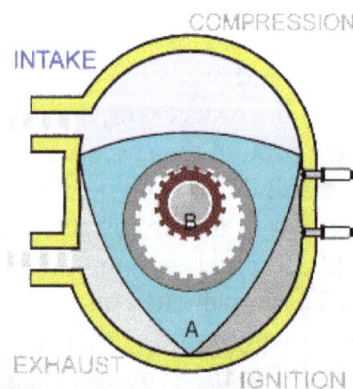

The Wankel KKM motorcycle: The "A" marks one of the three apices of the rotor. The "B" marks the eccentric shaft and the white portion is the lobe of the eccentric shaft. The shaft turns 3 times for each rotation of the rotor around the lobe and once for each orbital revolution around the eccentric shaft.

The central drive shaft, called the "eccentric shaft" or "E-shaft", passes through the center of the rotor and is supported by fixed bearings. The rotors ride on eccentrics (analogous to crankpins) integral to the eccentric shaft (analogous to a crankshaft). The rotors both rotate around the eccentrics and make orbital revolutions around the eccentric shaft. Seals at the corners of the rotor seal against the periphery of the housing, dividing it into three moving combustion chambers. The rotation of each rotor on its own axis is caused and controlled by a pair of synchronizing gears A fixed gear mounted on one side of the rotor housing engages a ring gear attached to the rotor and ensures the rotor moves exactly 1/3 turn for each turn of the eccentric shaft. The power output of the engine is not transmitted through the synchronizing gears. The force of gas pressure on the rotor (to a first approximation) goes directly to the center of the eccentric, part of the output shaft...

The easiest way to visualize the action of the engine in the animation at left is to look not at the rotor itself, but the cavity created between it and the housing. The Wankel engine is actually a variable-volume progressing-cavity system. Thus, there are three cavities per housing, all repeating the same cycle. Points A and B on the rotor and E-shaft turn at different speeds—point B circles three times as often as point A does, so that one full orbit of the rotor equates to three turns of the E-shaft.

As the rotor rotates orbitally revolving, each side of the rotor is brought closer to and then away from the wall of the housing, compressing and expanding the combustion chamber like the strokes of a piston in a reciprocating piston engine. The power vector of the combustion stage goes through the center of the offset lobe.

While a four-stroke piston engine completes one combustion stroke per cylinder for every two rotations of the crankshaft (that is, one-half power stroke per crankshaft rotation per cylinder), each combustion chamber in the Wankel generates one combustion stroke per driveshaft rotation, i.e. one power stroke per rotor orbital revolution and three power strokes per rotor rotation. Thus, the power output of a Wankel engine is generally higher than that of a four-stroke piston engine of similar engine displacement in a similar state of tune; and higher than that of a four-stroke piston engine of similar physical dimensions and weight.

Wankel engines generally can sustain much higher engine revolutions than reciprocating engines of similar power output. This is due to the smoothness inherent in circular motion, and the absence of highly stressed parts such as crankshafts, camshafts or connecting rods. Eccentric shafts do not have the stress related contours of crankshafts. The maximum revolutions of a rotary engine is limited by tooth load on the synchronizing gears. Hardened steel gears are used for extended operation above 7000 or 8000 rpm. Mazda Wankel engines in auto racing are operated above 10,000 rpm. In aircraft they are used conservatively, up to 6500 or 7500 rpm. However, as gas pressure participates in seal efficiency, racing a Wankel engine at high rpm under no load conditions can destroy the engine.

National agencies that tax automobiles according to displacement and regulatory bodies in automobile racing variously consider the Wankel engine to be equivalent to a four-stroke piston engine of 1.5 to 2 times the displacement. Some racing series have banned the Wankel altogether.

Engineering

Felix Wankel managed to overcome most of the problems that made previous rotary engines fail by developing a configuration with vane seals that had a tip radius equal to the amount of "oversize" of the rotor housing form, as compared to the theoretical epitrochoid, to minimize radial apex seal motion plus introducing a cylindrical gas-loaded apex pin which abutted all sealing elements to seal around the three planes at each rotor apex.

Apex seals, left NSU Ro 80 Serie and Research and right Mazda 12A and 13B

Left Mazda old L10A camber axial cooling, middle Audi NSU EA871 axial water cooling only hot bow, right Diamond Engines Wankel radial cooling only in the hot bow

In the early days, special, dedicated production machines had to be built for different housing dimensional arrangements. However, patents such as U.S. Patent 3,824,746, G. J. Watt, 1974, for a "Wankel Engine Cylinder Generating Machine", U.S. Patent 3,916,738, "Apparatus for machining and/or treatment of trochoidal surfaces" and U.S. Patent 3,964,367, "Device for machining trochoidal inner walls", and others, solved the issue.

Rotary engines have a problem not found in reciprocating four-stroke engines in that the block housing has intake, compression, combustion, and exhaust occurring at fixed locations around the housing. In contrast, reciprocating engines perform these four strokes in one chamber, so that extremes of "freezing" intake and "flaming" exhaust are averaged and shielded by a boundary layer from overheating working parts. The use of Heat Pipes in an Air Cooled Wankel was proposed by the University of Florida to overcome this uneven heating of the block housing. Pre-heating of certain housing sections with exhaust gas improved performance and fuel economy, also reducing wear and emissions.

The boundary layer shields and the oil film act as thermal insulation, leading to a low temperature of the lubricating film (maximum ~200 °C/400 °F) on a water-cooled Wankel engine. This gives a more constant surface temperature. The temperature around the spark plug is about the same as the temperature in the combustion chamber of a reciprocating engine. With circumferential or axial flow cooling, the temperature difference remains tolerable.

During research in the 1950s and 1960s problems arose. For a while, engineers were faced with what they called "chatter marks" and "devil's scratch" in the inner epitrochoid surface. They discovered that the origin was in the apex seals reaching a resonating vibration, and solved the problem by reducing the thickness and weight of apex seals. Scratches disappeared after the introduction of more compatible materials for seals and housing coatings. Another early problem of the build-up of cracks in the stator surface near the plug hole was eliminated by installing the spark plugs in a separate metal insert/ copper sleeve in the housing instead of plug being screwed directly into the block housing. Toyota found that substituting a glow-plug for the leading site spark plug improved low rpm, part load, specific fuel consumption by 7% and also emissions and idle. A later alternative solution to spark plug boss cooling was provided with a variable coolant velocity scheme for water-cooled rotaries, which has had widespread use, being patented by Curtiss-Wright, with the last-listed for better air-cooled engine spark plug boss cooling. These approaches did not require a high conductivity copper insert, but did not preclude its use. Ford tested a rotary engine with the plugs placed in the side plates, instead of the usual placement in the housing working surface (CA 1036073, 1978).

Four-stroke reciprocating engines are less suitable for use with hydrogen fuel. The hydrogen can misfire on hot parts like the exhaust valve and spark plugs. Another problem concerns the hydrogenate attack on the lubricating film in reciprocating engines. In a Wankel engine, this problem is circumvented by using a ceramic apex seal against a ceramic surface; there is no oil film to suffer hydrogenate attack. The piston shell must be lubricated and cooled with oil. This substantially increases the lubricating oil consumption in a four-stroke hydrogen engine.

Increasing the displacement and power of a rotary engine by adding more rotors to a basic design is simple, but a limit may exist in the number of rotors, as power output is channeled through the last rotor shaft, with all the stresses of the whole engine present at this point. For engines with more than two rotors, the approach of coupling two bi-rotor sets by a serrate coupling between the two rotor sets has been tested successfully.

SPARCS in the United Kingdom found that idle stability and economy was obtained by supplying an ignitable mix to only one rotor in a multi rotor engine in a forced-air cooled rotor, similar to the Norton designs.

Materials

Unlike a piston engine, where the cylinder is heated by the combustion process and then cooled by the incoming charge, Wankel rotor housings are constantly heated on one side and cooled on the other, leading to high local temperatures and unequal thermal expansion. While this places high demands on the materials used, the simplicity of the Wankel makes it easier to use alternative materials, such as exotic alloys and ceramics. With water cooling in a radial or axial flow direction, with the hot water from the hot bow heating the cold bow, the thermal expansion remains tolerable; top engine temperature has been reduced to 129 °C, with a maximum temperature difference between engine parts of 18 °C by the use of heat pipes around the housing and in side plates as a cooling means.

Among the alloys cited for Wankel housing use are A-132, Inconel 625, and 356 treated to T6 hardness. Several materials have been used for plating the housing working surface, Nikasil being one. Citroen, Mercedes-Benz, Ford, A P Grazen and others applied for patents in this field. For the apex seals, the choice of materials has evolved along with the experience gained, from carbon alloys, to steel, ferrotic, and other materials. The combination between housing plating and apex and side seals materials was determined experimentally, to obtain the best duration of both seals and housing cover. For the shaft, steel alloys with little deformation on load are preferred, the use of Maraging steel has been proposed for this.

Lead is a solid lubricant with leaded gasoline linked to a reduced wear of seals and housings. Leaded gasoline was the predominant type available in the first years of the Wankel engine's development.

The first engines had the oil supply calculated with consideration of gasoline's lubricating qualities. Leaded gasoline was phased out, with Wankel engines needing an increased mix of oil in the gasoline to provide lubrication to critical engine parts. Experienced users advise, even in engines with electronic fuel injection, adding at least 1% of oil directly to gasoline as a safety measure in case the pump supplying oil to combustion chamber related parts fails or sucks in air. A SAE paper by David Garside extensively describes Norton's choices of materials and cooling fins.

Several approaches involving solid lubricants were tested, and even the addition of MoS2, one cc per liter of fuel is advised (LiquiMoly). Many engineers agree that the addition of oil to gasoline as in old two-stroke engines is a safer approach for engine reliability than an oil pump injecting into the intake system or directly to the parts requiring lubrication. A combined oil-in-fuel plus oil metering pump is always possible.

Sealing

Early engine designs had a high incidence of sealing loss, both between the rotor and the housing and also between the various pieces making up the housing. Also, in earlier model Wankel engines, carbon particles could become trapped between the seal and the casing, jamming the engine and requiring a partial rebuild. It was common for very early Mazda engines to require rebuilding after 50,000 miles (80,000 km). Further sealing problems arose from the uneven thermal distribution within the housings causing distortion and loss of sealing and compression. This thermal distortion also caused uneven wear between the apex seal and the rotor housing, evident on higher mileage engines. The problem was exacerbated when the engine was stressed before reaching operating temperature. However, Mazda rotary engines solved these initial problems. Current engines have nearly 100 seal-related parts.

The problem of clearance for hot rotor apexes passing between the axially closer side housings in the cooler intake lobe areas was dealt with by using an axial rotor pilot radially inboard of the oils seals, plus improved inertia oil cooling of the rotor interior (C-W US 3261542, C. Jones, 5/8/63, US 3176915, M. Bentele, C. Jones. A.H. Raye. 7/2/62), and slightly "crowned" apex seals (different height in the center and in the extremes of seal).

Modern Wankel engines have fully sealed mainshaft cases. Many engines do not require oil changes, as the oil is not contaminated by the combustion process.

Fuel Economy and Emissions

The shape of the Wankel combustion chamber is more resistant to preignition operating on lower-octane rating gasoline than a comparable piston engine. The combustion chamber shape may also lead to relatively incomplete combustion of the air-fuel charge. This would result in a larger amount of unburned hydrocarbons released into the exhaust. The exhaust is, however, relatively low in NOx emissions, as combustion temperatures are lower than in other engines, and also because of some inherent exhaust gas recirculation (EGR) in early engines. Sir Harry Ricardo showed in the 1920s that for every 1% increase in the proportion of exhaust gas in the admission mix, there is a 7 °C (45 °F) reduction in flame temperature. This allowed Mazda to meet the United States Clean Air Act of 1970 in 1973, with a simple and inexpensive "thermal reactor", which is an enlarged chamber in the exhaust manifold. By decreasing the air-fuel ratio, unburned hydrocarbons (HC) in the exhaust would support combustion in the thermal reactor. Piston-engine cars required expensive catalytic converters to deal with both unburned hydrocarbons and NOx emissions. This inexpensive solution improved fuel consumption, which was already a weak point for the Wankel engine, at the same time that the oil crisis of 1973 raised the price of gasoline. Toyota discovered that injection of air into the exhaust port zone improved fuel economy and emissions. The best results were obtained with holes in the side plates, doing it in the exhaust duct had no noticeable influence. The use of a three-stage catalysts, with air supplied in the middle, as for 2-Stroke piston engines, also proved good.

Mazda improved the fuel efficiency of the thermal reactor system by 40% by the time of introduction of the RX-7 in 1978. However, Mazda eventually shifted to the catalytic converter system. According to the Curtiss-Wright research, the factor that controls the amount of unburned HC in the exhaust is the rotor surface temperature, with higher temperatures producing less HC. Curtiss-Wright showed also that the rotor can be widened, keeping the rest of engine's architecture unchanged, thus reducing friction losses and increasing displacement and power output. The limiting factor for this widening being mechanical considerations, especially shaft deflection at high rotative speeds. Quenching is the dominant source of HC at high speeds, and leakage at low speeds.

Automobile Wankel rotary engines are capable of high-speed operation. However, it was shown that an early opening of the intake port, longer intake ducts, and a greater rotor eccentricity can increase torque at low rpm. The shape and positioning of the recess in the rotor, which forms most of the combustion chamber, influences emissions and fuel economy. The results in terms of fuel economy and exhaust emissions varies depending on the shape of the combustion recess which is determined by the placement of spark plugs per chamber of an individual engine.

Mazda's RX-8 car with the Renesis engine, fuel economy met California State requirements, including California's low emissions vehicle (LEV) standards. This was achieved by a number of innovations. The exhaust ports, which in earlier Mazda rotaries were located in the rotor housings, were moved to the sides of the combustion chamber. This solved the problem of the earlier ash buildup in the engine, and thermal distortion problems of side intake and exhaust ports. A scraper seal was added in the rotor sides, and by use of some ceramic-made parts in the engine. This approach allowed Mazda to eliminate overlap between intake and exhaust port openings, while simultaneously increasing the exhaust port area. The side port trapped the unburned fuel in the chamber, decreased the oil consumption, and improved the combustion stability in the low-speed and light load range. The HC emissions from the side exhaust port Wankel engine are 35–50% less than those from the peripheral exhaust port Wankel engine, because of near zero intake and exhaust port opening overlap. Peripheral ported rotary engines have a better mean effective pressure, especially at high rpm and with a rectangular shaped intake port. However, the RX-8 was not improved to meet EuroV emission regulations and was discontinued in 2012.

Mazda is still continuing development of the next generation of Wankel engines. The company is researching engine laser ignition, eliminating conventional spark plugs, and direct fuel injection and HCCI ignition. This leads to a greater rotor eccentricity, equating to a longer stroke in a reciprocating engine, for better elasticity and low revolutions per minute torque. Research by T Kohno proved that installing a glow-plug in the combustion chamber improved 7% part load and low revolutions per minute fuel economy. These innovations promise to improve fuel consumption and emissions. To improve fuel efficiency further, Mazda is looking at using the Wankel as a range extender in series-hybrid cars announcing a prototype, the Mazda2 EV, for press evaluation in November 2013. This configuration improves fuel efficiency and emissions. As a further advantage, running a Wankel engine at a constant speed gives greater engine life. Keeping to a near constant, or narrow band, of revolutions eliminates, or vastly reduces, many of the disadvantages of the Wankel engine.

Laser Ignition

As the rotor's apex seals pass over the spark plug hole, compressed charge can be lost from the charge chamber to the exhaust chamber, entailing fuel in the exhaust, reducing efficiency, and giving high emissions. The spark plug needs to be located outside the combustion chamber to enable the apex of the rotor to sweep past. These points may be overcome by using laser ignition, eliminating traditional spark plugs, which may give a narrow slit in the motor housing the rotor apex seals can fully cover with no loss of compression from adjacent chambers. The laser plug can fire its spark through the narrow slit. Laser spark plugs can fire deep into the combustion chamber using multiple sparks. Direct fuel injection of which the Wankel engine is suited, combined with laser ignition in single or multiple laser plugs, has shown to enhance the motor even further reducing the disadvantages.

Homogeneous Charge Compression Ignition (HCCI)

Homogeneous charge compression ignition (HCCI) is where the fuel/air intake is a pre-mixed lean air-fuel mixture then compressed to the point of auto-ignition. Electronic spark ignition is eliminated. Gasoline engines combine homogeneous charge (HC) with spark ignition (SI), abbreviated as HCSI. Diesel engines combine stratified charge (SC) with compression ignition (CI), abbreviated as SCCI. HCCI engines achieve gasoline engine-like emissions with compression ignition engine-like efficiency. HCCI engines achieve low levels of nitrogen oxide emissions (NO x) without a catalytic converter. However, unburned hydrocarbon and carbon monoxide emissions still require treatment to reach automotive emission regulations.

Mazda have undertaken research on HCCI ignition for its SkyActiv-R rotary engine project using research from its SkyActiv Generation 2 program. A constraint of rotary engines is the need to locate the spark plug outside the combustion chamber to enable the rotor to sweep past. Mazda confirmed this has been solved in the SkyActiv-R project. Rotaries generally have high compression ratios leaning the design to ease of homogeneous charge compression ignition (HCCI) adoption.

Advantages

NSU Wankel Spider, the first line of cars sold with a rotor Wankel engine

Mazda Cosmo, the first series two rotor Wankel engine sports car

Prime advantages of the Wankel engine are:

- A far higher power to weight ratio than a piston engine (it is approximately one third of the weight of a piston engine of equivalent power output)

- It is approximately one third of the size of a piston engine of equivalent power output

- No reciprocating parts

- Able to reach higher revolutions per minute than a piston engine

- Operates with almost no vibration

- Not prone to engine-knock

- Cheaper to mass-produce as the engine contains fewer parts

- Superior breathing, filling the combustion charge in 270 degrees of mainshaft rotation rather than 180 degrees in a piston engine

- Supplies torques for about two thirds of the combustion cycle rather than one quarter for a piston engine

- Wider speed range gives greater adaptability

- It can use fuels of wider octane ratings

- Does not suffer from "scale effect" to limit its size

- On some Wankel engines the sump oil remains uncontaminated by the combustion process requiring no oil changes. The oil in the mainshaft is totally sealed from the combustion process. The oil for Apex seals and crankcase lubrication is separate. In piston engines the crankcase oil is contaminated by combustion blow-by through the piston rings.

Wankel engines are considerably lighter and simpler, containing far fewer moving parts than piston engines of equivalent power output. Valves or complex valve trains are eliminated by using simple ports cut into the walls of the rotor housing. Since the rotor rides directly on a large bearing on the output shaft, there are no connecting rods and no crankshaft. The elimination of reciprocating mass and the elimination of the most highly stressed and failure prone parts of piston engines gives the Wankel engine high reliability, a smoother flow of power, and a high power-to-weight ratio.

The surface-to-volume-ratio in the moving combustion chamber is so complex that a direct comparison cannot be made between a reciprocating piston engine and a Wankel engine. The flow velocity and the heat losses behave quite differently. Surface temperatures behave absolutely differently; the film of oil in the Wankel engine acts as insulation. Engines with a higher compression ratio have a worse surface-to-volume ratio. The surface-to-volume ratio of a reciprocating piston diesel engine is much poorer than a reciprocating piston gasoline engine, however diesel engines have a higher efficiency factor. Hence, comparing power outputs is a realistic metric. A reciprocating piston engine with equal power to a Wankel will be approximately twice the displacement. When comparing the power-to-weight ratio, physical size or physical weight to a similar power output piston engine, the Wankel is superior.

A four-stroke cylinder produces a power stroke only every other rotation of the crankshaft, with three strokes being pumping losses. This doubles the real surface-to-volume ratio for the four-stroke reciprocating piston engine and the displacement increased. The Wankel, therefore, has higher volumetric efficiency and lower pumping losses through the absence of choking valves. Because of the quasi-overlap of the power strokes that cause the smoothness of the engine and the avoidance of the four-stroke cycle as in a reciprocating engine, the Wankel engine is very quick to react to power increase changes giving a quick delivery of power when the demand arises, especially at higher rpm's. This difference is more pronounced when compared to four-cylinder reciprocating engines and less pronounced when compared to higher cylinder counts.

In addition to the removal of internal reciprocating stresses by virtue of the complete removal of reciprocating internal parts typically found in a piston engine, the Wankel engine is constructed with an iron rotor within a housing made of aluminium, which has a greater coefficient of thermal expansion. This ensures that even a severely overheated Wankel engine cannot seize, as would be likely to occur in an overheated piston engine. This is a substantial safety benefit of use in aircraft. In addition, the absence of valves and valve trains again increases safety. GM tested an Iron Rotor and Iron Housing in their prototype Wankel engines, that worked at higher temperatures with lower specific fuel consumption.

A further advantage of the Wankel engine for use in aircraft is that a Wankel engine generally has a smaller frontal area than a piston engine of equivalent power, allowing a more aerodynamic nose to be designed around the engine. A cascading advantage is the smaller size and less weight of the Wankel engine also allows for savings in airframe construction costs, compared to piston engines of comparable power output.

Wankel engines that operate within their original design parameters are almost immune to catastrophic failure. A Wankel engine that loses compression, cooling or oil pressure will lose a large amount of power and fail over a short period of time. It will, however, usually continue to produce some power during that time, allowing for a safer landing when used in aircraft. Piston engines under the same circumstances are prone to seizing or breaking parts that almost certainly results in catastrophic failure of the engine and instant full loss of power. For this reason, Wankel engines are very well suited to snowmobiles, which often take users into remote places where a failure could result in frostbite or death, and aircraft, where abrupt failure is likely to lead to a crash or forced landing in a remote place.

From the combustion chamber shape and features, the fuel ON requirements of Wankel engines are lower than in reciprocating piston engines. The maximum road octane number requirements were 82 for a peripheral intake port wankel engine, and less than 70 for a side inlet port engine. From the point of view of oil refiners this may be an industrial advantage in fuel production costs.

Due to a 50% longer stroke duration than a reciprocating four-cycle engine, there is more time to complete the combustion. This leads to greater suitability for direct fuel injection and stratified charge operation. A Wankel rotary engine has stronger flows of air-fuel mixture and a longer operating cycle than a reciprocating engine, realizing concomitantly thorough mixing of hydrogen and air. The result is a homogeneous mixture, and no hot spots in the engine, which is crucial for hydrogen combustion.

Disadvantages

Many of the disadvantages are in ongoing research with some advances greatly reducing negative aspects of the engine. However, the current disadvantages of the Wankel engine in production are the following:

- Rotor sealing. This is still a minor problem as the engine housing has vastly different temperatures in each separate chamber section. The different expansion coefficients of the materials gives a far from perfect sealing. Additionally, both sides of the seals are being exposed to fuel, and the design does not allow for a dedicated lubrication system, as in two-stroke engines. In comparison, a piston engine has all functions of a cycle in the same chamber giving a more stable temperature for piston rings to act against; additionally, only one side of the piston in a (four-stroke) piston engine is being exposed to fuel, allowing for oil to lubricate the cylinders from the other side. To overcome the differences in temperatures between different regions of housing and side and intermediary plates, and the associated thermal dilatation inequities, the use of a heat pipe, transporting heat from the hot to the cold parts of engine, has been shown to reduce, in a small displacement, charge cooled rotor, air-cooled housing wankel engine, the maximal engine temperature from 231 °C to 129 °C, and the maximum difference from a hotter to a colder region of engine, from 159 °C to 18 °C.

- Apex seal lifting. Centrifugal force pushes the apex seal onto the housing surface forming a firm seal. Gaps can develop between the apex seal and troichoid housing in light-load operation when imbalances in centrifugal force and gas pressure occur. In low engine-rpm ranges, or under low-load conditions, gas pressure in the combustion chamber can cause the seal to lift off the surface, resulting in combustion gas leaking into the next chamber. Mazda has identified this problem and has developed a solution. By changing the shape of the troichoid housing, the seals remain flush to the housing. This points to using the engine at sustained higher revolutions eliminating apex seal lift off, in applications such as an electricity generator. In vehicles this leads to series-hybrid applications of the engine.

- Slow combustion. The combustion is slow as the combustion chamber is long, thin, and moving. The trailing side of the combustion chamber naturally produces a "squeeze stream" that prevents the flame from reaching the chamber trailing edge. Fuel injection in which fuel is injected towards the leading edge of the combustion chamber can minimize the amount of unburnt fuel in the exhaust. Kawasaki proposed a triangular tail extension of the plug hole, pointing to the combustion chamber trailing side to solve this.

- Bad fuel economy. This is due to seals leakages, and the "difficult" shape of the combustion chamber, with poor combustion behavior and mean effective pressure at part load, low rpm. Meeting the emissions regulations requirements sometimes mandates a fuel-air ratio that is not the best for fuel economy. Acceleration and deceleration as in direct drive average driving conditions also affect fuel economy. Running the engine at a constant speed and load eliminates excess fuel consumption.

- Poor emissions. As unburnt fuel is in the exhaust stream, emissions requirements are difficult to meet. This problem may be overcome by implementing direct fuel injection into the combustion chamber. The Freedom Motors Rotapower Wankel engine, which is not yet in production, met the ultra low California emissions standards. The Mazda Renesis engine, with both intake and exhaust side ports, suppressed the loss of unburned mix to exhaust formerly induced by port overlap.

Although in two dimensions the seal system of a Wankel looks to be even simpler than that of a corresponding multi-cylinder piston engine, in three dimensions the opposite is true. As well as the rotor apex seals evident in the conceptual diagram, the rotor must also seal against the chamber ends.

Piston rings are not perfect seals: each has a gap to allow for expansion. The sealing at the Wankel apexes is less critical, as leakage is between adjacent chambers on adjacent strokes of the cycle, rather than to the crankcase. Although sealing has improved over the years, the less than effective sealing of the Wankel, which is mostly due to lack of lubrication, is still a factor reducing its efficiency. Comparison tests have shown that the Mazda rotary powered RX-8 sports car may use more fuel than a heavier vehicle powered by larger displacement V-8 engines for similar performance results.

The fuel-air mixture cannot be pre-stored as there are consecutive intake cycles. The Wankel engine has a 50% longer stroke duration than a piston engine. The four Otto cycles last 1080° for a Wankel engine (three revolutions of the output shaft) versus 720° for a four-stroke reciprocating piston engine, but the four strokes are still the same proportion of the total.

There are various methods of calculating the engine displacement of a Wankel. The Japanese regulations for calculating displacements for engine ratings use the volume displacement of one rotor face only, and the auto industry commonly accepts this method as the standard for calculating the displacement of a rotary. When compared by specific output, however, the convention results in large imbalances in favor of the Wankel motor, an early approach was rating displacement of each rotor as two times the chamber.

Wankel rotary engine and piston engine displacement and corresponding power output can more accurately be compared by displacement per revolution of the eccentric shaft. A calculation of this form dictates that a two-rotor Wankel displacing 654 cc per face will have a displacement of 1.3 liters per every rotation of the eccentric shaft (only two total faces, one face per rotor going through a full power stroke) and 2.6 liters after two revolutions (four total faces, two faces per rotor going through a full power stroke). The results are directly comparable to a 2.6-liter piston engine with an even number of cylinders in a conventional firing order, which will likewise displace 1.3 liters through its power stroke after one revolution of the crankshaft, and 2.6 liters through its power strokes after two revolutions of the crankshaft. A Wankel rotary engine is still a four-stroke engine and pumping losses from non-power strokes still apply, but the absence of throttling valves and a 50% longer stroke duration result in a significantly lower pumping loss compared to a four-stroke reciprocating piston engine. Measuring a Wankel rotary engine in this way more accurately explains its specific output, as the volume of its air fuel mixture put through a complete power stroke per revolution is directly responsible for torque and thus power produced.

The trailing side of the rotary engine's combustion chamber develops a squeeze stream which pushes back the flamefront. With the conventional one or two-spark-plug system and homogenous mixture, this squeeze stream prevents the flame from propagating to the combustion chamber's trailing side in the mid and high engine speed ranges, Mazda engineers described the full process. Kawasaki addressed this problem in their US patent US 3848574, and Toyota obtained a 7% economy improvement by placing a glow-plug in the leading site and using Reed-Valves in intake ducts. This poor combustion in the trailing side of chamber is one of the reasons why there is more carbon monoxide and unburnt hydrocarbons in a Wankel's exhaust stream. A side-port exhaust, as is used in the Mazda Renesis, avoids one of the causes of this because the unburned mixture cannot escape. The Mazda 26B avoided this issue through a three spark-plug ignition system. (At the 24 Hours of Le Mans endurance race in 1991 the 26B had significantly lower fuel consumption than the competing reciprocating piston engines. All competitors had the same amount of fuel available due to the Le Mans limited fuel quantity rule.)

A peripheral intake port gives the highest mean effective pressure, however, side intake porting produces a more steady idle, as it helps to prevent blow-back of burned gases into the intake ducts which cause "misfirings": alternating cycles where the mixture ignites and fails to ignite; peripheral porting (PP) gives the best mean effective pressure throughout the rpm range, but PP was linked also to worse idle stability and part-load performance. Early work from Toyota led to the addition of a fresh air supply to the exhaust port and proved also that a Reed-valve in the intake port or ducts improved the low rpm and partial load performance of wankel engines, by preventing blow-back of exhaust gas into the intake port and ducts, and reducing the misfiring-inducing high EGR, at the cost of a small loss of power at top rpm; this is according to David W. Garside, the developer of the Norton rotary engine, who proposed that an earlier opening of the intake port before top dead center (TDC) and longer intake ducts improved low rpm torque and elasticity of wankel engines, also described in Kenichi Yamamoto's books. Elasticity is also improved with a greater rotor eccentricity, analogous to a longer stroke in a reciprocating engine. Wankel engines operate better with a low pressure exhaust system, higher exhaust back pressure reducing mean effective pressure, more severely in peripheral intake port engines. The Mazda RX-8 Renesis engine improved performance by doubling the exhaust port area respect to earlier designs, and there is specific work about the effect of intake and exhaust piping configuration on wankel engines performance.

All Mazda-made Wankel rotaries, including the Renesis found in the RX-8, burn a small quantity of oil by design, metered into the combustion chamber to preserve the apex seals. Owners must periodically add small amounts of oil, thereby increasing running costs. Some sources (rotaryeng. net) claim that better results come with the use of an oil-in-fuel mixture rather than an oil metering pump. Liquid cooled engines require a mineral multigrade oil for cold starts, and wankel engines need a warm-up time before full load operation as reciprocating engines do. All engines exhibit oil loss, however the rotary engine is engineered with a sealed motor, unlike a piston engine that has a film of oil that splashes on the walls of the cylinder to lubricate them, hence an oil "control" ring. No-oil-loss engines have been developed, eliminating much of the oil lubrication problems.

Applications

Automobile Racing

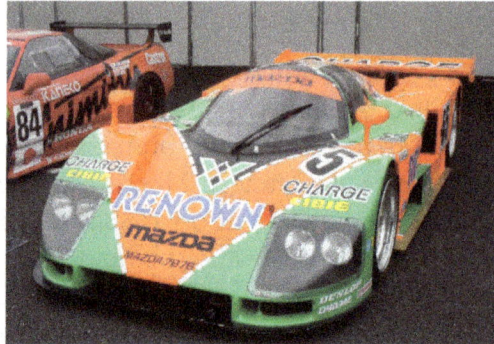

Mazda 787B

In the racing world, Mazda has had substantial success with two-rotor, three-rotor, and four-rotor cars. Private racers have also had considerable success with stock and modified Mazda Wankel-engine cars.

The Sigma MC74 powered by a Mazda 12A engine was the first engine and only team from outside Western Europe or the United States to finish the entire 24 hours of the 24 Hours of Le Mans race, in 1974. Mazda is the only team from outside Western Europe or the United States to have won Le Mans outright and the only non-piston engine ever to win Le Mans, which the company accomplished in 1991 with their four-rotor 787B (2,622 cc or 160 cu in—actual displacement, rated by FIA formula at 4,708 cc or 287 cu in).

Formula Mazda Racing features open-wheel race cars with Mazda Wankel engines, adaptable to both oval tracks and road courses, on several levels of competition. Since 1991, the professionally organized Star Mazda Series has been the most popular format for sponsors, spectators, and upward bound drivers. The engines are all built by one engine builder, certified to produce the prescribed power, and sealed to discourage tampering. They are in a relatively mild state of racing tune, so that they are extremely reliable and can go years between motor rebuilds.

The Malibu Grand Prix chain, similar in concept to commercial recreational kart racing tracks, operates several venues in the United States where a customer can purchase several laps around a track in a vehicle very similar to open wheel racing vehicles, but powered by a small Curtiss-Wright rotary engine.

In engines having more than two rotors, or two rotor race engines intended for high-rpm use, a multi-piece eccentric shaft may be used, allowing additional bearings between rotors. While this approach does increase the complexity of the eccentric shaft design, it has been used successfully in the Mazda's production three-rotor 20B-REW engine, as well as many low volume production race engines. The C-111-2 4 Rotor Mercedes-Benz eccentric shaft for the KE Serie 70, Typ DB M950 KE409 is made in one piece. Mercedes-Benz used split bearings.

Motorcycle Engines

Norton Interpol2 prototype

The small size and attractive power to weight ratio of the Wankel engine appealed to motorcycle manufacturers. The first Wankel-engined motorcycle was the 1960 'IFA/MZ KKM 175W' built by German motorcycle manufacturer MZ licensed from NSU.

In 1972, Yamaha introduced the RZ201 at the Tokyo Motor Show, a prototype with a Wankel engine, weighing 220 kg and producing 60 hp from a twin-rotor 660 cc engine (US patent N3964448). In 1972 Kawasaki presented its two-rotor Kawasaki X99 rotary engine prototype (US patents N 3848574 &3991722). Both Yamaha and Kawasaki claimed to have solved the problems in early Wankels, namely: poor fuel economy, high exhaust emissions, and poor engine longevity, but neither prototype reached production.

In 1974 Hercules produced W-2000 Wankel motorcycles, but low production numbers meant the project was unprofitable, and production ceased in 1977.

From 1975 to 1976, Suzuki produced its RE5 single-rotor Wankel motorcycle. It was a complex design, with both liquid cooling and oil cooling, and multiple lubrication and carburetor systems. It worked well and was smooth, but being rather heavy and having a modest 62 bhp power output, it did not sell well.

Dutch motorcycle importer and manufacturer Van Veen produced small quantities of their dual-rotor Wankel-engined OCR-1000 between 1978 and 1980, using surplus Comotor engines.

In the early 1980s, using earlier work at BSA, Norton produced the air-cooled twin-rotor Classic, followed by the liquid-cooled Commander and the Interpol2 (a police version). Subsequent Norton Wankel bikes included the Norton F1, F1 Sports, RC588, Norton RCW588, and NRS588. Norton has proposed a new 588 cc twin-rotor model called the "NRV588" and a 700 cc version called the "NRV700". A former mechanic at Norton, Brian Crighton, started developing his own rotary engined motorcycles line named "Roton", whose products won several local Australian races.

Despite successes in racing, no motorcycles powered by Wankel engines have been produced for sale to the general public for road use since 1992.

The two different design approaches, taken by Suzuki and BSA may usefully be compared. Even before Suzuki produced the RE5, in Birmingham BSA's research engineer David Garside, was developing a twin-rotor Wankel motorcycle. BSA's collapse put a halt on development, but Garside's machine eventually reached production as the Norton Classic.

Wankel engines run very hot on the ignition and exhaust side of the engine's trochoid chamber, whereas the intake and compression parts are cooler. Suzuki opted for a complicated oil-cooling and water cooling system, with Garside reasoning that provided the power did not exceed 80 bhp, air-cooling would suffice. Garside cooled the interior of the rotors with filtered ram-air. This very hot air was cooled in a plenum contained within the semi-monocoque frame and afterwards, once mixed with fuel, fed into the engine. This air was quite oily after running through the interior of the rotors, and thus was used to lubricate the rotor tips. The exhaust pipes become very hot, with Suzuki opting for a finned exhaust manifold, twin-skinned exhausted pipes with cooling grilles, heatproof pipe wrappings and silencers with heat shields. Garside simply tucked the pipes out of harm's way under the engine, where heat would dissipate in the breeze of the vehicle's forward motion. Suzuki opted for complicated multi-stage carburation, whilst Garside choose simple carburetors. Suzuki had three lube systems, whilst Garside had a single total-loss oil injection system which was fed to both the main bearings and the intake manifolds. Suzuki chose a single rotor that was fairly smooth, however with rough patches at 4,000 rpm; Garside opted for a turbine-smooth twin-rotor motor. Suzuki mounted the massive rotor high in the frame; Garside put his rotors as low as possible to lower the center of gravity of the motorcycle.

Although it was said to handle well, The result was that the Suzuki was heavy, overcomplicated, expensive to manufacture, and (at 62 bhp) a little short on power. Garside's design was simpler, smoother, lighter and (at 80 bhp) significantly more powerful.

Aircraft Engines

Diamond DA20 with Diamond Engines Wankel

Sikorsky Cypher Unmanned aerial vehicle (UAV) powered with a UEL AR801 Wankel engine

ARV Super2 with the British MidWest AE110 twin-rotor Wankel engine

In principle, a Wankel engine should be ideal for light aircraft, as it is light, compact, almost vibrationless and has a high power-to-weight ratio. Further aviation benefits of a Wankel engine include:

1. Rotors cannot seize, since rotor casings expand more than rotors;

2. A Wankel engine is less prone to the serious condition known as "engine-knock", which can destroy the plane's engines in mid-flight.

3. A Wankel is not susceptible to "shock-cooling" during descent;

4. A Wankel does not require an enriched mixture for cooling at high power;

5. Having no reciprocating parts, there is less vulnerability to damage when the engines revolves higher than the designed maximum running operation. The limit to the revolutions is the strength of the main bearings.

Unlike the case with some cars and motorcycles, a Wankel aero-engine will be sufficiently warm before full power is asked of it because of the time taken for pre-flight checks. A Wankel aero-engine spends most of its operational time at high power outputs, with little idling. This makes ideal the use of peripheral ports. An advantage is that modular engines with more than two rotors are feasible. If icing of any intake tracts is an issue, there is plenty of waste engine heat available to prevent icing.

The first Wankel rotary-engine aircraft was the experimental Lockheed Q-Star civilian version of the United States Army's reconnaissance QT-2, basically a powered Schweizer sailplane, in the late 1960s. The plane was powered by a 185 hp (138 kW) Curtiss-Wright RC2-60 Wankel rotary engine; the same engine model was also flown in a Cessna Cardinal and other airplanes and a helicopter. In Germany in the mid-1970s, a pusher ducted fan airplane powered by a modified NSU multi-rotor Wankel engine was developed in both civilian and military versions, Fanliner and Fantrainer.

In roughly the same timeframe as the first experiments with full-scale aircraft powered with Wankel engines, model aircraft-sized versions were pioneered by a combine of the well-known Japanese O.S. Engines firm and the then-extant German Graupner aeromodeling products firm, under license from NSU/Auto-Union. By 1968 the first prototype air-cooled, single-rotor glow plug-ignition, methanol fueled 4.9 cm^3 displacement OS/Graupner model Wankel engine was running, and was produced in at least two differing versions from 1970 to the present day, solely by the O.S. firm since Graupner's demise in 2012.

Aircraft Wankels have been taken up with the advantages over other engines being exploited. Wankels are increasingly being found in roles where the compact size, high power-to-weight ratio and quiet operation is important, notably in drones and unmanned aerial vehicles. Many companies and hobbyists adapt Mazda rotary engines (taken from automobiles) to aircraft use; others, including Wankel GmbH itself, manufacture Wankel rotary engines dedicated for this purpose. One such use are the "Rotapower" engines in the Moller Skycar M400. Another example of purpose built aircraft rotaries are Austro Engine's 55 hp (40.4 kW) AE50R (certified) and 75 hp (55 kW) AE75R (under development) both appr. 2 hp/kg.

Wankel engines are also becoming increasingly popular in homebuilt experimental aircraft, such as the ARV Super2 which can be re-engined with the British MidWest AE series aero-engine. Most are Mazda 12A and 13B automobile engines, converted to aviation use. This is a very cost-effective alternative to certified aircraft engines, providing engines ranging from 100 to 300 horsepower (220 kW) at a fraction of the cost of traditional engines. These conversions first took place in the early 1970s. With a number of these engines mounted on aircraft, as of 10 December 2006 the National Transportation Safety Board has only seven reports of incidents involving aircraft with Mazda engines, and none of these were a failure due to design or manufacturing flaws.

Peter Garrison, contributing editor for *Flying* magazine, has said that "In my opinion, however, the most promising engine for aviation use is the Mazda rotary." Mazdas have indeed worked well when converted for use in homebuilt aircraft. However, the real challenge in aviation is producing FAA-certified alternatives to the standard reciprocating engines that power most small general aviation aircraft. Mistral Engines, based in Switzerland, developed purpose-built rotaries for factory and retrofit installations on certified production aircraft. The G-190 and G-230-TS rotary engines were already flying in the experimental market, and Mistral Engines hoped for FAA and JAA certification by 2011. As of June 2010, G-300 rotary engine development ceased, with the company citing a need for cash flow to complete development.

Mistral claims to have overcome the challenges of fuel consumption inherent in the rotary, at least to the extent that the engines are demonstrating specific fuel consumption within a few points of reciprocating engines of similar displacement. While fuel burn is still marginally higher than traditional engines, it is outweighed by other beneficial factors.

At the price of increased complication for a high pressure diesel type injection system, fuel consumption in the same range as small pre-chamber automotive and industrial diesels has been demonstrated with Curtiss-Wright's stratified charge multi-fuel engines, while preserving the aforementioned Wankel rotary advantages Unlike a piston and overhead valve engine, there are no valves which can float at higher rpm causing loss of performance. The Wankel is a more effective design at high revolutions with no reciprocating parts, far fewer moving parts and no cylinder head.

The French company Citroën had developed Wankel powered RE-2 (fr) helicopter in the 1970s.

The British company Rotron, who specialise in unmanned aerial vehicle (UAV) applications of Wankel engines have designed and built a unit to operate on heavy fuel for NATO purposes. The prime innovation is flame propagation, ensuring the flame burns smoothly across the whole combustion chamber. The fuel is pre-heated to 98 degrees Celsius before injection into the combustion chamber. Four spark plugs are utilised aligned in two pairs. Two spark plugs ignite the fuel charge at the front of the rotor as it moves into the combustion section of the housing. As the rotor moves the fuel charge, the second two fire a fraction of second behind the first pair of plugs igniting near the rear of the rotor at the back of the fuel charge. The drive shaft is water cooled which has a cooling effect on the internals of the rotor. Cooling water also flows around the external of the engine through a gap in the housing, hence cooling the engine from outside and inside eliminating hot spots.

Since Wankel engines operate at a relatively high rotational speed, at 6'000 rpm of output shaft, the Rotor makes only 2'000 turns, with relatively low torque, propeller driven aircraft must use a propeller speed reduction unit to maintain propellers within the designed speed range. Experimental aircraft with Wankel engines use propeller speed reduction units, for instance the MidWest twin-rotor engine has a 2.95:1 reduction gearbox. The rotational shaft speed of a Wankel engine is high compared to reciprocating piston designs. Only the eccentric shaft spins fast, while the rotors turn at exactly one-third of the shaft speed. If the shaft is spinning at 7,500 rpm, the rotors are turning at a much slower 2,500 rpm.

Pratt & Whitney Rocketdyne have been commissioned by DARPA to develop a diesel Wankel engine for use in a prototype VTOL flying car called the "Transformer". The engine, based on an earlier unmanned aerial vehicle Wankel diesel concept called "Endurocore", plans to utilize Wankel rotors of varying sizes on a shared eccentric shaft to increase efficiency. The engine is claimed to be a 'full-compression, full-expansion, diesel-cycle engine'. An October 28, 2010 patent from Pratt & Whitney Rocketdyne, describes a Wankel engine superficially similar to Rolls-Royce's earlier prototype that required an external air compressor to achieve high enough compression for diesel-cycle combustion. The design differs from Rolls-Royce's diesel Wankel mainly by proposing an injector both in the exhaust passage between the combustor rotor and expansion rotor stages, and an injector in the expansion rotor's expansion chamber, for 'afterburning'.

The sailplane manufacturer Schleicher uses Wankel engines in their self-launching models ASK-21 Mi, ASH-26E, ASH-25 M/Mi, ASH-30 Mi, ASH-31 Mi, ASW-22 BLE, and ASG-32 Mi.

In 2013 e-Go aeroplanes, based in Cambridge, United Kingdom, announced that their new single-seater canard aircraft, the winner of a design competition to meet the new UK single-seat deregulated category, will be powered by a Wankel engine from Rotron Power, a specialist manufacturer of advanced rotary engines for unmanned aeronautical vehicle (UAV) applications. The first sale was 2016. The aircraft is expected to deliver 100 kts cruise speed from a 30 hp Wankel engine, with a fuel economy of 75 mpg using standard motor gasoline (MOGAS), developing 22 kW (30 hp).

The DA36 E-Star, an aircraft designed by Siemens, Diamond Aircraft and EADS, employs a series hybrid powertrain with the propeller being turned by a Siemens 70 kW (94 hp) electric motor. The aim is to reduce fuel consumption and emissions by up to 25 percent. An onboard 40 hp (30 kW) Austro Engines Wankel rotary engine and generator provides the electricity. A propeller speed reduction unit is eliminated. The electric motor uses electricity stored in batteries, with the generator engine off, to take off and climb reducing sound emissions. The series-hybrid powertrain using the Wankel engine reduces the weight of the plane by 100 kilograms from its predecessor. The DA36 E-Star first flew in June 2013, making this the first ever flight of a series-hybrid powertrain. Diamond Aircraft state that the technology using Wankel engines is scalable to a 100-seat aircraft.

Range Extender

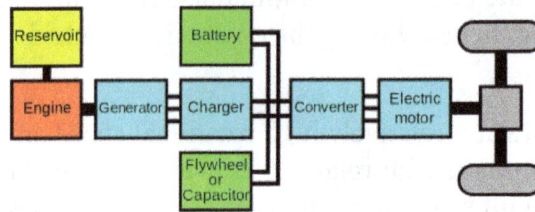

Structure of a series-hybrid vehicle. The grey square represents a differential gear. An alternative arrangement (not shown) is to have electric motors at two or four wheels.

Mazda2 EV prototype

Due to the compact size and the high power to weight ratio of a Wankel engine, a number have been proposed for electric vehicles as range extenders to supplement when electric battery levels are low, with a number of concept cars incorporating a series hybrid powertrain arrangement. A Wankel engine used only in a generator setup has packaging and weight distribution advantages, maximizing interior passenger and luggage space when used in a vehicle. The engine/generator may be at one end of the vehicle with the electric driving motors at the other connected only by thin light cables. In 2010 Audi revealed a prototype series-hybrid electric car, the A1 e-tron, that incorporated a small 250 cc Wankel engine running at 5,000 rpm recharging the car's batteries as needed, and providing electricity directly to the electric driving motor. In 2010 FEV Inc revealed that in their prototype electric version of the Fiat 500 a Wankel engine would also be used as a range extender. Valmet Automotive of Finland in 2013 revealed a prototype incorporating a Wankel powered series-hybrid powertrain car named the EVA, utilizing an engine manufactured by the German company Wankel SuperTec.

Mazda of Japan ceased production of direct drive Wankel engines in their model range in 2012, leaving the motor industry world-wide with no production cars using the engine. The company is continuing development of the next generation of their Wankel engines, the SkyActiv-R with a new rear wheel drive sports car model announced in October 2015 although with no launch date given. Mazda states that the SkyActiv-R solves the three key issues with previous rotary engines: fuel economy, emissions and reliability. Mr Takashi Yamanouchi, the global CEO of Mazda stated, "The rotary engine has very good dynamic performance, but it's not so good on economy when you accelerate and decelerate. However, with a range extender you can use a rotary engine at a constant 2,000rpm, at its most efficient. It's compact, too." No Wankel engine in this arrangement, as yet has made it into production vehicles or planes. However, in November 2013 Mazda announced a series-hybrid prototype car to the motoring press, the Mazda2 EV using a Wankel engine as a range extender. The engine is a tiny, almost inaudible, single-rotor 330cc unit generating 30 bhp at 4,500rpm maintaining a continuous electric output of 20 kW. The engine is located under the rear luggage floor.

Other Uses

Small Wankel engines are being found increasingly in other applications, such as go-karts, personal water craft and auxiliary power units for aircraft. Kawasaki patented also a mixture cooled rotary engine (US patent 3991722). Yanmar Diesel and Dolmar-Sachs had a rotary engine chain saw (SAE paper 760642) and outboard boat engines, and the French Outils Wolf, a Wankel rotary engine powered lawnmower (Rotondor), with the rotor in a horizontal position and no seals in the down side, for production costs savings. The Graupner/O.S. 49-PI is a 1.27 hp (947 W) 5 cc Wankel engine for model airplane use which has been in production essentially unchanged since 1970; even with a large muffler, the entire package weighs only 380 grams (13.4 ounces).

UEL UAV-741 Wankel engine for a UAV

The simplicity of the Wankel makes it well-suited for mini, micro, and micro-mini engine designs. The Microelectromechanical systems (MEMS) Rotary Engine Lab at the University of California, Berkeley, has previously undertaken research towards the development of Wankel engines of down to 1 mm in diameter with displacements less than 0.1 cc. Materials include silicon and motive power includes compressed air. The goal of such research was to eventually develop an internal combustion engine with the ability to deliver 100 milliwatts of electrical power; with the engine itself serving as the rotor of the generator, with magnets built into the engine rotor itself. Development of the miniature Wankel engine stopped at UC Berkeley at the end of the DARPA contract. Miniature Wankel engines struggled to maintain compression due to sealing problems, similar to problems observed in the large scale versions. In addition, miniature engines suffer from an adverse surface to volume ratio causing excess heat losses; the relatively large surface area of the combustion chamber walls transfers away what little heat is generated in the small combustion volume resulting in quenching and low efficiency.

Ingersoll-Rand built the largest ever Wankel engine which was available between 1975 and 1985 producing 1,100 hp (820 kW) with two rotors. A one rotor version was available producing 550 hp (410 kW). The displacement per rotor was 41 liters with each rotor approximately one meter in diameter. The engine was derived from a previous, unsuccessful Curtiss-Wright design, which failed because of a well-known problem with all internal combustion engines: the fixed speed at which the flame front travels limits the distance combustion can travel from the point of ignition in a given time, and thereby limiting the maximum size of the cylinder or rotor chamber which can be used. This problem was solved by limiting the engine speed to only 1200 rpm and the use of natural gas as fuel; this was particularly well chosen, since one of the major uses of the engine was to drive compressors on natural gas pipelines.

Yanmar Diesel of Japan produced some small, charge-cooled rotor rotary engines for uses such as chainsaws and outboard engines, some of their contributions are the LDR (rotor recess in the leading edge of combustion chamber) engines having better exhaust emissions profiles, and that reed-valve controlled intake ports improved part-load and low rpm performance.

In 1971 and 1972, Arctic Cat produced snowmobiles powered by Sachs KM 914 303 cc and KC-24 294 cc Wankel Engine made in Germany.

In the early 1970s Johnson and other brands sold Snowmobiles powered by 35 or 45 HP Wankel engines designed and built by OMC.

Aixro of Germany produces and sells a 294 cc per chamber charge-cooled rotor and liquid-cooled housings kart engines, other makers are: Wankel AG, Cubewano, Rotron, Precision Technology USA.

Non-internal Combustion

In addition for use as an internal combustion engine, the basic Wankel design has also been used for gas compressors, and superchargers for internal combustion engines, but in these cases, although the design still offers advantages in reliability, the basic advantages of the Wankel in size and weight over the four-stroke internal combustion engine are irrelevant. In a design using a Wankel supercharger on a Wankel engine, the supercharger is twice the size of the engine.

The Wankel design is used in the seat belt pre-tensioner system of some Mercedes-Benz and Volkswagen cars. When the deceleration sensors sense a potential crash, small explosive cartridges are triggered electrically and the resulting pressurized gas feeds into tiny Wankel engines which rotate to take up the slack in the seat belt systems, anchoring the driver and passengers firmly in the seat before a collision.

Gunpowder Engine

A gunpowder engine, also known as an explosion engine or Huygens' engine, is a type of internal combustion engine using gunpowder as its fuel. The concept was first explored during the 1600s, most notably by famous Dutch polymath Christiaan Huygens. George Cayley also experimented with the design in the early 1800s as an aircraft engine, and claims to have made models that worked for a short time. There is also a persistent claim that conventional carboretted gasoline engine can be run on gunpowder, but no examples of a successful conversion can be documented.

Earliest Mentions

The earliest references to a gunpowder engine appear to be those of Samuel Morland in 1661. This consists solely of a letter of patent written by King Charles the Second that was received at

Whitehall on 11 December 1661. No other information about this "engine" remains, but the description involves the use of vacuum to draw water.

The next known reference is by Jean de Hautefeuille in 1678, suggested as a solution to the problem of raising water from the Seine to supply Versailles. He presented two ideas, one using the vacuum like Morland's idea, and a second that used a U-shaped tube with water in one side and air in the other. When the gunpowder was lit in the air-filled side, the rise in pressure would drive the water up the other side.

Like early steam engine designs, these engines used the air or vacuum created by gunpowder to directly lift the water. There were no mechanical parts in the manner of modern engines, which translate the power in the gas pressure into any needed mechanical form.

Huygens and Papin

In 1671, Denis Papin was given a job at the Academy of the Royal Library in Paris, where he worked under the Curator of Experiments, Christiaan Huygens. Huygens set Papin to the task of carrying out a research effort on air and vacuum, at that time a matter of widespread international study. As part of the experiments, Papin measured the force of a small amount of gunpowder lit in small iron and copper vessels. Papin published an account of all of these experiments in 1674 in *New experiments on the vacuum, with a description of the machines used for making them.*

Papin moved to London shortly after publication, and from then on was more involved in the development of steam. Although his developments pointed the way towards the early steam engine, Papin himself became more interested in the latent heat of steam and developed the "steam digester", the first pressure cooker. He also conceived of a number of devices using air pressure as a working fluid, include a series of fountains, pumps, and similar devices.

In spite of there being no further examples of particle work on the part of Papin, he did carry on a continued correspondence with Gottfried Wilhelm Leibniz on this and other topics. Leibniz tried to interest Papin in further development throughout, at one point noting "Yet I would well counsel [you], Monsieur, to undertake more considerable things which would force everyone to give their approbation and would truly change the state of things. The two items of binding together the pneumatic machine and gunpowder and applying the force of fire to vehicles would truly be of this nature" Papin replied that he had constructed a small model of a paddle-wheel boat, but the type of engine is not stated.

Huygens' Engine

Huygens, however, became interested in the mechanical power of the vacuum, and the possibility of using gunpowder to produce one. In 1678 he outlined a gunpowder engine consisting of a vertical tube containing a piston. Gunpowder was inserted into the tube and lit through a small hole at the base, like a cannon. The expanding gasses would drive the piston up the tube until the reached a point near the top. Here, the piston uncovered holes in the tube that allowed any remaining hot gasses to escape. The weight of the piston and the vacuum formed by the cooling gasses in the now-closed cylinder drew the piston back into the tube, lifting a test mass to provide power.

According to sources, a single example of this sort of engine was built in 1678 or 79 using a cannon as the cylinder. The cylinder was held down to a base where the gunpowder sat, making it a breech loading design. The gasses escaped via two leather tubes attached at the top of the barrel. When the piston reached them the gasses blew the tubes open, and when the pressure fell, gravity pulled the leather down causing the tubes droop to the side of the cylinder, sealing the holes.

Huygens' presented a paper on his invention in 1680, *A New Motive Power by Means of Gunpowder and Air*. By 1682, the device had successfully shown that a dram (1/16th of an ounce) of gunpowder, in a cylinder seven or eight feet high and fifteen or eighteen inches in diameter, could raise seven or eight boys (or about 1,100 pounds) into the air, who held the end of the rope. However, there is considerable debate in modern sources as to whether or not the engine could have been built. Sealing the piston within the cylinder proved to be a very difficult problem in modern recreations.

From that point, few mentions of early gunpowder engines are found. The use of steam, especially after the introduction of the atmospheric engine in 1712, captured all further development effort.

Cayley

As part of his investigations of powered flight, George Cayley was concerned about the low power-to-weight ratio of steam engines, complaining that "the steam engine has hither proved to weighty and cumbrous for most purposes of locomotion." He took up development of a new engine design starting in 1807, and quickly settled on a gunpowder engines as the preferred solution, noting "Being in want of a simple & light first mover on a small scale for the purpose of some preparatory experiments on aerial navigation, I constructed one in which the force of gunpowder & the heat evolved by its explosion, acting upon a quantity of common air, was employed."

His notebooks show a design of considerable improvement over those of Huygens and similar. In Cayley's design, two cylinders were arranged one over the other, the lower acting as a combustion chamber, and the upper containing a piston. A small charge of gunpowder was introduced into the bottom of the lower cylinder and lit by a hot rod heated by candles. The expanded gasses pushed the piston up, and this energy was captured in a large bow, in effect, drawing the bow back as if readying to fire an arrow. The bowstring pushed the piston rod back down as the gasses escaped and cooled, completing the cycle.

In a later version, Cayley attempted to solve the problem of continual cycling. In this version, the combustion chamber was removed to a separate cylinder placed to the side of the power cylinder. Gunpowder was stored in the upper portion of this chamber, and small amounts were metered out to fall into the combustion area below. The hot gasses were then piped out of the combustion area into the power cylinder. This consisted of two pistons on a common piston rod, with the gasses flowing into alternate sides of the cylinder to form a double-acting engine.

In a letter, Cayley stated that he had constructed one of these designs (although which is not mentioned), but also stated that it did not work very well. Over time he designed several flying machines using the engine, but no larger working model appears to have been attempted.

Paine and Others

Thomas Paine introduced an entirely new type of engine design, one that bore more resemblance to a water wheel than a conventional engine. In Paine's engine, a series of cup-like combustion chambers were arranged around a wheel. As the wheel turned, each cup received a small amount of gunpowder from a central container and was then lit.

The literature contains numerous other mentions of gunpowder engines, but it does not appear any were used operationally.

In Modern Engines

The idea that a conventional gasoline engine can be run on gunpowder is a persistent topic of discussion. It was taken up by MythBusters on Episode 63, and after a number of attempts it was considered "busted".

Small Engine

Electrical generator, with Generac Vanguard (Now Briggs Vanguard) engine and electric start

Rotary lawn mower, with a vertical shaft engine

A small engine is the general term for a wide range of small-displacement, low-powered internal combustion engines used to power lawn mowers, generators, concrete mixers and many other machines that require independent power sources. Most small engines are single-cylinder, with a few V-twin units. Although much less common, there have been small Wankel (rotary) engines manufactured for use on lawn mowers and other such equipment.

Power Range

The engines, which may be of two or four-stroke design, are small in both physical dimensions and power output, relative to larger automobile engines. Power output ranges from less than 1 to about 15 horsepower. The smallest of all are used in handheld garden machinery, such as string trimmers and chainsaws, which may be as small as 25 cc (2 cu in) piston displacement. The most common are four-stroke air-cooled single-cylinder engines running on either petrol or diesel.

Working Cycles

Engines for small machinery that must be hand-carried, such as string trimmers and chainsaws, are usually a two-stroke design, which is lighter for any given power output. However, two-stroke engines create a relatively large amount of air pollution and noise pollution, and so are beginning to be supplanted by four-stroke units.

Design

Compared to modern vehicle engines, small engines are relatively simple in design. Capital cost is usually more important than fuel economy, running costs or longevity, thus encouraging simple designs.

Valves

Valves are most commonly actuated overhead, in a OHV configuration. An example of one of the best known and most popular designs is the Honda GX range, dating from the early 1980s. The sidevalve arrangement still persists, owing to its simplicity for both manufacture and basic maintenance. A few engines, such as the Honda GC series, now use overhead cam valvegear, driven by a timing belt. This engine also uses an unusual monobloc design where the cylinder head, block, and half the crankcase share the same casting, termed 'uniblock' by Honda.

Ignition

Electrical systems for these engines are usually simple and minimal, mostly to avoid the need for a battery. Many uses require the engines to be ready for immediate use, even after long standby storage and the risk of a flat battery. Most small engines are equipped with capacitor discharge ignition, that—like the magneto that preceded it—is an ignition system that does not require a battery.

Starting

Electric starting is available for small engines and is found primarily on high-feature garden machinery and larger generators particularly where there is already a complex electrical system and there may be a need for auto-starting on demand. However, a self-retracting rope-pull mechanism called a recoil starter is the predominant method of starting small engines; it does not require a battery to power a starter motor, nor an alternator to keep a battery charged. Before the invention of the recoil starter, a notched pulley was attached to the engine's flywheel; the operator would manually wind a rope around the pulley then jerk the rope to rotate the engine so that it would start. Another starting method briefly popular in the 1960s was the "impulse" or "wind-up" starter. These were operated by winding a heavy spring by means of a rotating crank handle equipped with a sprag clutch, then releasing the spring's tension by means of a lever or knob so that it would spin the engine. These were discontinued when safety problems became apparent: it was possible to leave the starter wound up and ready to start the engine unintentionally, even long after the crank was wound up, if the release were jarred.

Fuel System

Fuel systems are usually simple, at least for petrol engines. The fuel tank may be placed above a float-type carburettor so that gravity is adequate to bring fuel to the carburettor, avoiding the need for a fuel pump. Alternatively, the tank may be placed below a carburettor which uses engine vacuum or crankcase pressure pulsations to lift fuel into the carburetor. Float-type carburettors can only be used when the engine is reliably vertical and without excessive vibration.

Governor

Most small engines have a governor to regulate and maintain the engine speed as constantly as possible despite changing load. Some engines also have an adjustable throttle by which the operator or the machine itself can alter the engine speed. Generally, the nominal "throttle control" does not directly control the opening of the carburetor throttle, but rather increases or decreases spring tension on the governor, which in turn regulates the engine speed higher or lower.

Crankshaft

Small engines may be configured with the crankshaft horizontal or vertical, according to the intended application. Vertical axis engines were developed for rotary lawnmowers, but the size of this large market has encouraged a supply of cheap engines and they are now used for other purposes too, such as generators.

History

Small engines date back to the early days of internal combustion engines. The first of a recognisably modern form date from the advent of motorcycles, where the engines were made by the same manufacturers.

Major manufacturers today

- Briggs & Stratton

- Hatz Diesel

- Honda

- Kawasaki

- Kohler

- Ryobi

- Subaru Industrial Power Products

- Tanaka (2-stroke)

- Vortex engines (Italy) 2-stroke cart engines

- Yanmar (small diesel engines)

Major Chinese manufacturers

- Launtop

- Loncin

- HuaSheng

Past manufacturers

- Clinton

- Cushman

- J.A.P.

- Tecumseh

- Villiers

Free-piston Engine

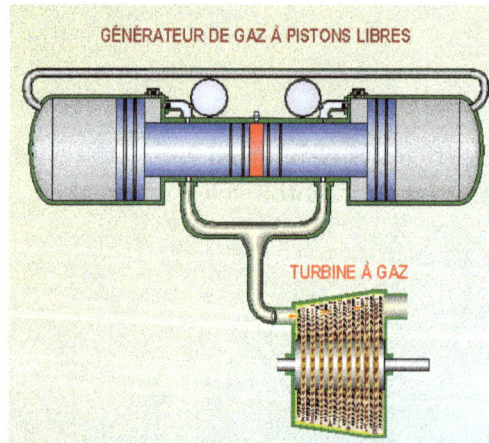

Free-piston engine used as a gas generator to drive a turbine

A free-piston engine is a linear, 'crankless' internal combustion engine, in which the piston motion is not controlled by a crankshaft but determined by the interaction of forces from the combustion chamber gases, a rebound device (e.g., a piston in a closed cylinder) and a load device (e.g. a gas compressor or a linear alternator).

The purpose of all such piston engines is to generate power. In the free-piston engine, this power is not delivered to a crankshaft but is instead extracted through either exhaust gas pressure driving a turbine, through driving a linear load such as an air compressor for pneumatic power, or by incorporating a linear alternator directly into the pistons to produce electrical power.

The basic configuration of free-piston engines is commonly known as single piston, dual piston or opposed pistons, referring to the number of combustion cylinders. The free-piston engine is usually restricted to the two-stroke operating principle, since a power stroke is required every fore-and-aft cycle. However, a split cycle four-stroke version has been patented, GB2480461 (A) published 2011-11-23.

First Generation

The modern free-piston engine was proposed by R.P. Pescara and the original application was a single piston air compressor. Pescara set up the *Bureau Technique Pescara* to develop free-piston engines and Robert Huber was technical director of the Bureau from 1924 to 1962.

The engine concept was a topic of much interest in the period 1930-1960, and a number of commercially available units were developed. These first generation free-piston engines were without exception opposed piston engines, in which the two pistons were mechanically linked to ensure symmetric motion. The free-piston engines provided some advantages over conventional technology, including compactness and a vibration-free design.

Air Compressors

The first successful application of the free-piston engine concept was as air compressors. In these engines, air compressor cylinders were coupled to the moving pistons, often in a multi-stage configuration. Some of these engines utilised the air remaining in the compressor cylinders to return the piston, thereby eliminating the need for a rebound device.

Free-piston air compressors were in use among others by the German Navy, and had the advantages of high efficiency, compactness and low noise and vibration.

Gas Generators

After the success of the free-piston air compressor, a number of industrial research groups started the development of free-piston gas generators. In these engines there is no load device coupled to the engine itself, but the power is extracted from an exhaust turbine. (The only load for the engine is supercharging the inlet air.)

A number of free-piston gas generators were developed, and such units were in widespread use in large-scale applications such as stationary and marine powerplants. Attempts were made to use free-piston gas generators for vehicle propulsion (e.g. in gas turbine locomotives) but without success.

Modern Applications

Modern applications of the free-piston engine concept include hydraulic engines, aimed for off-highway vehicles, and free-piston engine generators, aimed for use with hybrid electric vehicles.

Hydraulic

These engines are commonly of the single piston type, with the hydraulic cylinder acting as both load and rebound device using a hydraulic control system. This gives the unit high operational flexibility. Excellent part load performance has been reported.

Generators

Free-piston linear generators that eliminate a heavy crankshaft with electrical coils in the piston and cylinder walls are being investigated by multiple research groups for use in hybrid electric vehicles as range extenders. The first free piston generator was patented in 1934. Examples include the Stelzer engine and the Free Piston Power Pack manufactured by Pempek Systems based on a German patent. A single piston Free-piston linear generator was demonstrated in 2013 at the German Aerospace Center (Deutsches Zentrum für Luft- und Raumfahrt; DLR).

These engines are mainly of the dual piston type, giving a compact unit with high power-to-weight ratio. A challenge with this design is to find an electric motor with sufficiently low weight. Control challenges in the form of high cycle-to-cycle variations were reported for dual piston engines.

In June 2014 Toyota announced a prototype Free Piston Engine Linear Generator (FPEG). As the piston is forced downward during its power stroke it passes through windings in the cylinder to generate a burst of three-phase AC electricity. The piston generates electricity on both strokes, reducing piston dead losses. The generator operates on a two-stroke cycle, using hydraulically activated exhaust poppet valves, gasoline direct injection and electronically operated valves. The engine is easily modified to operate under various fuels including hydrogen, natural gas, ethanol, gasoline and diesel. A two-cylinder FPEG is inherently balanced.

Toyota claims a thermal-efficiency rating of 42% in continuous use, greatly exceeding today's average of 25-30%. Toyota demonstrated a 24 inch long by 2.5 inch in diameter unit producing 15 hp (greater than 11 kW).

Features

The operational characteristics of free-piston engines differ from those of conventional, crankshaft engines. The main difference is due to the piston motion not being restricted by a crankshaft in the free-piston engine, leading to the potentially valuable feature of variable compression ratio. This does, however, also present a control challenge, since the position of the dead centres must be accurately controlled in order to ensure fuel ignition and efficient combustion, and to avoid excessive in-cylinder pressures or, worse, the piston hitting the cylinder head.

Advantages

Potential advantages of the free-piston concept include

- Simple design with few moving parts, giving a compact engine with low maintenance costs and reduced frictional losses.

- The operational flexibility through the variable compression ratio allows operation optimisation for all operating conditions and multi-fuel operation. The free-piston engine is further well suited for homogeneous charge compression ignition (HCCI) operation.

- High piston speed around top dead centre (TDC) and a fast power stroke expansion enhances fuel-air mixing and reduces the time available for heat transfer losses and the formation of temperature-dependent emissions such as nitrogen oxides (NOx).

Challenges

The main challenge for the free-piston engine is engine control, which can only be said to be fully solved for single piston hydraulic free-piston engines. Issues such as the influence of cycle-to-cycle variations in the combustion process and engine performance during transient operation in dual piston engines are topics that need further investigation. Crankshaft engines can connect traditional accessories such as alternator, oil pump, fuel pump, cooling system, starter etc.

Rotational movement to spin conventional automobile engine accessories such as alternators, air conditioner compressors, power steering pumps, and anti-pollution devices could be captured from a turbine situated in the exhaust stream.

Opposing Piston Engine

Most free piston engines are of the opposed piston type with a single central combustion chamber. A variation is the Opposing piston engine which has two separate combustion chambers. An example is the Stelzer engine.

Recent Developments

In the 21st century, research continues into free-piston engines and patents have been published in many countries. In the UK, Newcastle University is undertaking research into free-piston engines.

A new kind of the free-piston engine, a Free-piston linear generator is being developed by the German aerospace center.

Jet Engine

A jet engine is a reaction engine discharging a fast-moving jet that generates thrust by jet propulsion. This broad definition includes turbojets, turbofans, rocket engines, ramjets, and pulse jets. In general, jet engines are combustion engines.

In common parlance, the term *jet engine* loosely refers to an internal combustion airbreathing jet engine. These typically feature a rotating air compressor powered by a turbine, with the leftover power providing thrust via a propelling nozzle — this process is known as the Brayton thermodynamic cycle. Jet aircraft use such engines for long-distance travel. Early jet aircraft used turbojet engines which were relatively inefficient for subsonic flight. Modern subsonic jet aircraft usually use more complex high-bypass turbofan engines. These engines offer high speed and greater fuel efficiency than piston and propeller aeroengines over long distances.

A Pratt & Whitney F100 turbofan engine for the F-15 Eagle being tested in the hush house at Florida Air National Guard base. The tunnel behind the engine muffles noise and allows exhaust to escape

U.S. Air Force F-15E Strike Eagles

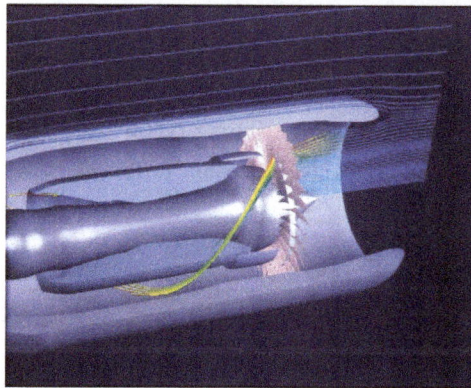

Simulation of a low-bypass turbofan's airflow.

Jet engine airflow during take-off.

The thrust of a typical jetliner engine went from 5,000 lbf (22,000 N) (de Havilland Ghost turbojet) in the 1950s to 115,000 lbf (510,000 N) (General Electric GE90 turbofan) in the 1990s, and their reliability went from 40 in-flight shutdowns per 100,000 engine flight hours to less than one in the late 1990s. This, combined with greatly decreased fuel consumption, permitted routine transatlantic flight by twin-engined airliners by the turn of the century, where before a similar journey would have required multiple fuel stops.

History

Jet engines date back to the invention of the aeolipile before the first century AD. This device directed steam power through two nozzles to cause a sphere to spin rapidly on its axis. So far as is known, it did not supply mechanical power and the potential practical applications of this invention did not receive recognition. Instead, it was seen as a curiosity.

Jet propulsion only gained practical applications with the invention of the gunpowder-powered rocket by the Chinese in the 13th century as a type of firework, and gradually progressed to propel formidable weaponry. However, although very powerful, at reasonable flight speeds rockets are very inefficient and so jet propulsion technology stalled for hundreds of years.

The earliest attempts at airbreathing jet engines were hybrid designs in which an external power source first compressed air, which was then mixed with fuel and burned for jet thrust. In one such system, called a *thermojet* by Secondo Campini but more commonly, motorjet, the air was compressed by a fan driven by a conventional piston engine. Examples of this type of design were the Caproni Campini N.1, and the Japanese Tsu-11 engine intended to power Ohka kamikaze planes towards the end of World War II. None were entirely successful and the N.1 ended up being slower than the same design with a traditional engine and propeller combination.

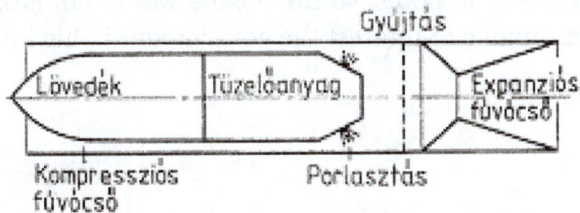

Albert Fonó's ramjet-cannonball from 1915

Even before the start of World War II, engineers were beginning to realize that engines driving propellers were self-limiting in terms of the maximum performance which could be attained; the limit was due to issues related to propeller efficiency, which declined as blade tips approached the speed of sound. If aircraft performance were ever to increase beyond such a barrier, a way would have to be found to use a different propulsion mechanism. This was the motivation behind the development of the gas turbine engine, commonly called a "jet" engine.

The key to a practical jet engine was the gas turbine, used to extract energy from the engine itself to drive the compressor. The gas turbine was not an idea developed in the 1930s: the patent for a stationary turbine was granted to John Barber in England in 1791. The first gas turbine to successfully run self-sustaining was built in 1903 by Norwegian engineer Ægidius Elling. Limitations in design and practical engineering and metallurgy prevented such engines reaching manufacture. The main problems were safety, reliability, weight and, especially, sustained operation.

The first patent for using a gas turbine to power an aircraft was filed in 1921 by Frenchman Maxime Guillaume. His engine was an axial-flow turbojet. Alan Arnold Griffith published *An Aerodynamic Theory of Turbine Design* in 1926 leading to experimental work at the RAE.

The Whittle W.2/700 engine flew in the Gloster E.28/39, the first British aircraft to fly with a turbojet engine, and the Gloster Meteor

In 1928, RAF College Cranwell cadet Frank Whittle formally submitted his ideas for a turbojet to his superiors. In October 1929 he developed his ideas further. On 16 January 1930 in England, Whittle submitted his first patent (granted in 1932). The patent showed a two-stage axial compressor feeding a single-sided centrifugal compressor. Practical axial compressors were made possible by ideas from A.A.Griffith in a seminal paper in 1926 ("An Aerodynamic Theory of Turbine Design"). Whittle would later concentrate on the simpler centrifugal compressor only, for a variety of practical reasons. Whittle had his first engine running in April 1937. It was liquid-fuelled, and included a self-contained fuel pump. Whittle's team experienced near-panic when the engine would not stop, accelerating even after the fuel was switched off. It turned out that fuel had leaked into the engine and accumulated in pools, so the engine would not stop until all the leaked fuel had burned off. Whittle was unable to interest the government in his invention, and development continued at a slow pace.

Heinkel He 178, the world's first aircraft to fly purely on turbojet power

In 1935 Hans von Ohain started work on a similar design in Germany, initially unaware of Whittle's work.

Von Ohain's first device was strictly experimental and could run only under external power, but he was able to demonstrate the basic concept. Ohain was then introduced to Ernst Heinkel, one of the larger aircraft industrialists of the day, who immediately saw the promise of the design. Heinkel had recently purchased the Hirth engine company, and Ohain and his master machinist Max Hahn were set up there as a new division of the Hirth company. They had their first HeS 1 centrifugal engine running by September 1937. Unlike Whittle's design, Ohain used hydrogen as fuel, supplied under external pressure. Their subsequent designs culminated in the gasoline-fuelled HeS 3 of 5 kN (1,100 lbf), which was fitted to Heinkel's simple and compact He 178 airframe and flown by Erich Warsitz in the early morning of August 27, 1939, from Rostock-Marienehe aerodrome, an impressively short time for development. The He 178 was the world's first jet plane.

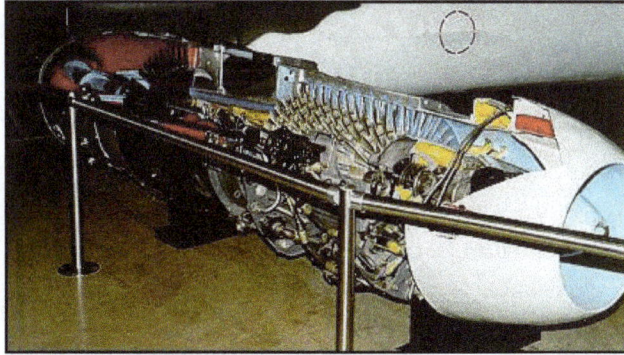
A cutaway of the Junkers Jumo 004 engine

Austrian Anselm Franz of Junkers' engine division (*Junkers Motoren* or "Jumo") introduced the axial-flow compressor in their jet engine. Jumo was assigned the next engine number in the RLM 109-0xx numbering sequence for gas turbine aircraft powerplants, "004", and the result was the Jumo 004 engine. After many lesser technical difficulties were solved, mass production of this engine started in 1944 as a powerplant for the world's first jet-fighter aircraft, the Messerschmitt Me 262 (and later the world's first jet-bomber aircraft, the Arado Ar 234). A variety of reasons conspired to delay the engine's availability, causing the fighter to arrive too late to improve Germany's position in World War II. Nonetheless, it will be remembered as the first use of jet engines in service.

Meanwhile, in Britain the Gloster E28/39 had its maiden flight on 15 May 1941 and the Gloster Meteor finally entered service with the RAF in July 1944. These were powered by turbojet engines from Power Jets Ltd., set up by Frank Whittle.

Following the end of the war the German jet aircraft and jet engines were extensively studied by the victorious allies and contributed to work on early Soviet and US jet fighters. The legacy of the axial-flow engine is seen in the fact that practically all jet engines on fixed-wing aircraft have had some inspiration from this design.

By the 1950s the jet engine was almost universal in combat aircraft, with the exception of cargo, liaison and other specialty types. By this point some of the British designs were already cleared for civilian use, and had appeared on early models like the de Havilland Comet and Avro Canada Jetliner. By the 1960s all large civilian aircraft were also jet powered, leaving the piston engine in low-cost niche roles such as cargo flights.

The efficiency of turbojet engines was still rather worse than piston engines, but by the 1970s, with the advent of high-bypass turbofan jet engines (an innovation not foreseen by the early commentators such as Edgar Buckingham, at high speeds and high altitudes that seemed absurd to them), fuel efficiency was about the same as the best piston and propeller engines.

Uses

A JT9D turbofan jet engine installed on a Boeing 747 aircraft.

Jet engines power jet aircraft, cruise missiles and unmanned aerial vehicles. In the form of rocket engines they power fireworks, model rocketry, spaceflight, and military missiles.

Jet engines have propelled high speed cars, particularly drag racers, with the all-time record held by a rocket car. A turbofan powered car, ThrustSSC, currently holds the land speed record.

Jet engine designs are frequently modified for non-aircraft applications, as industrial gas turbines or marine powerplants. These are used in electrical power generation, for powering water, natural gas, or oil pumps, and providing propulsion for ships and locomotives. Industrial gas turbines can create up to 50,000 shaft horsepower. Many of these engines are derived from older military turbojets such as the Pratt & Whitney J57 and J75 models. There is also a derivative of the P&W JT8D low-bypass turbofan that creates up to 35,000 HP.

Jet engines are also sometimes developed into, or share certain components such as engine cores, with turboshaft and turboprop engines, which are forms of gas turbine engines that are typically used to power helicopters and some propeller-driven aircraft..

Types

There are a large number of different types of jet engines, all of which achieve forward thrust from the principle of *jet propulsion.*

Airbreathing

Commonly aircraft are propelled by airbreathing jet engines. Most airbreathing jet engines that are in use are turbofan jet engines, which give good efficiency at speeds just below the speed of sound.

Turbine Powered

Gas turbines are rotary engines that extract energy from a flow of combustion gas. They have an upstream compressor coupled to a downstream turbine with a combustion chamber in-between. In aircraft engines, those three core components are often called the "gas generator." There are many different variations of gas turbines, but they all use a gas generator system of some type.

Turbojet

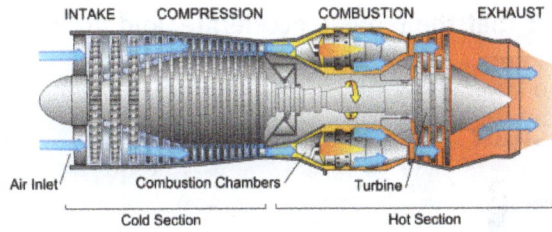

Turbojet engine

A turbojet engine is a gas turbine engine that works by compressing air with an inlet and a compressor (axial, centrifugal, or both), mixing fuel with the compressed air, burning the mixture in the combustor, and then passing the hot, high pressure air through a turbine and a nozzle. The compressor is powered by the turbine, which extracts energy from the expanding gas passing through it. The engine converts internal energy in the fuel to kinetic energy in the exhaust, producing thrust. All the air ingested by the inlet is passed through the compressor, combustor, and turbine, unlike the turbofan engine described below.

Turbofan

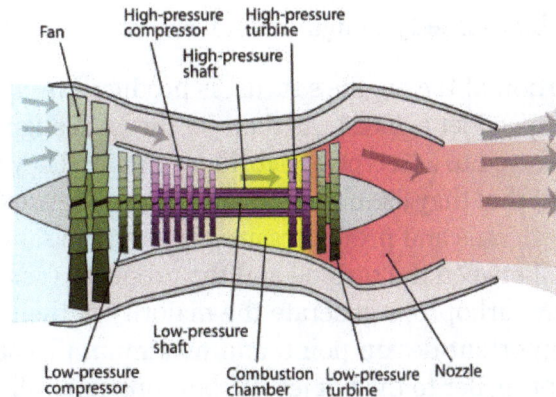

Schematic diagram illustrating the operation of a low-bypass turbofan engine.

A turbofan engine is a gas turbine engine that is very similar to a turbojet. Like a turbojet, it uses the gas generator core (compressor, combustor, turbine) to convert internal energy in fuel to kinetic energy in the exhaust. Turbofans differ from turbojets in that they have an additional component, a fan. Like the compressor, the fan is powered by the turbine section of the engine. Unlike the turbojet, some of the flow accelerated by the fan bypasses the gas generator core of the engine and is exhausted through a nozzle. The bypassed flow is at lower velocities, but a higher mass, making thrust produced by the fan more efficient than thrust produced by the core. Turbofans are generally more efficient than turbojets at subsonic speeds, but they have a larger frontal area which generates more drag.

There are two general types of turbofan engines, low-bypass and high-bypass. Low-bypass turbofans have a bypass ratio of around 2:1 or less, meaning that for each kilogram of air that passes through the core of the engine, two kilograms or less of air bypass the core. Low-bypass turbofans often use a mixed exhaust nozzle meaning that the bypassed flow and the core flow exit from the same nozzle. High-bypass turbofans have larger bypass ratios, sometimes on the order of 5:1 or 6:1. These turbofans can produce much more thrust than low-bypass turbofans or turbojets because of the large mass of air that the fan can accelerate, and are often more fuel efficient than low-bypass turbofans or turbojets.

Turboprop and Turboshaft

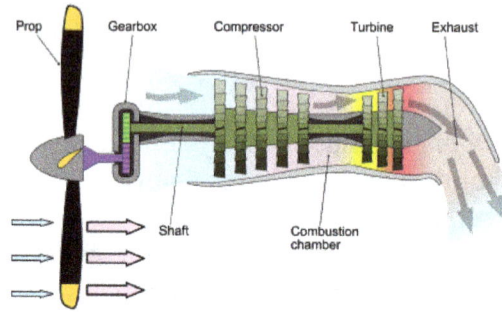

Turboprop engine

Turboprop engines are jet engine derivatives, still gas turbines, that extract work from the hot-exhaust jet to turn a rotating shaft, which is then used to produce thrust by some other means. While not strictly jet engines in that they rely on an auxiliary mechanism to produce thrust, turboprops are very similar to other turbine-based jet engines, and are often described as such.

In turboprop engines, a portion of the engine's thrust is produced by spinning a propeller, rather than relying solely on high-speed jet exhaust. As their jet thrust is augmented by a propeller, turboprops are occasionally referred to as a type of hybrid jet engine. They are quite similar to turbofans in many respects, except that they use a traditional propeller to provide the majority of thrust, rather than a ducted fan. Both fans and propellers are powered the same way, although most turboprops use gear-reduction between the turbine and the propeller (geared turbofans also feature gear reduction). While many turboprops generate the majority of their thrust with the propeller, the hot-jet exhaust is an important design point, and maximum thrust is obtained by matching thrust contributions of the propeller to the hot jet. Turboprops generally have better performance than turbojets or turbofans at low speeds where propeller efficiency is high, but become increasingly noisy and inefficient at high speeds.

Turboshaft engines are very similar to turboprops, differing in that nearly all energy in the exhaust is extracted to spin the rotating shaft, which is used to power machinery rather than a propeller, they therefore generate little to no jet thrust and are often used to power helicopters.

Propfan

A propfan engine (also called "unducted fan", "open rotor", or "ultra-high bypass") is a jet engine that uses its gas generator to power an exposed fan, similar to turboprop engines. Like turboprop engines, propfans generate most of their thrust from the propeller and not the exhaust jet. The primary difference between turboprop and propfan design is that the propeller blades on a propfan are highly swept to allow them to operate at speeds around Mach 0.8, which is competitive with modern commercial turbofans. These engines have the fuel efficiency advantages of turboprops with the performance capability of commercial turbofans. While significant research and testing (including flight testing) has been conducted on propfans, no propfan engines have entered production.

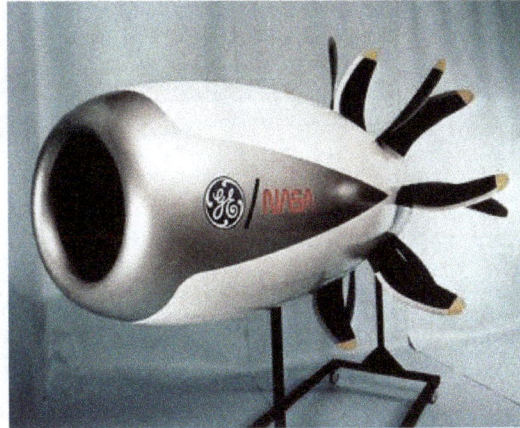
A propfan engine

Ram powered

Ram powered jet engines are airbreathing engines similar to gas turbine engines and they both follow the Brayton cycle. Gas turbine and ram powered engines differ, however, in how they compress the incoming airflow. Whereas gas turbine engines use axial or centrifugal compressors to compress incoming air, ram engines rely only on air compressed through the inlet or diffuser. Ram powered engines are considered the most simple type of air breathing jet engine because they can contain no moving parts.

Ramjet

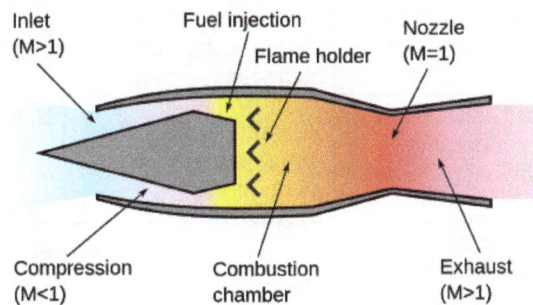
A schematic of a ramjet engine, where "M" is the Mach number of the airflow.

Ramjets are the most basic type of ram powered jet engines. They consist of three sections; an inlet to compress incoming air, a combustor to inject and combust fuel, and a nozzle to expel the hot gases and produce thrust. Ramjets require a relatively high speed to efficiently compress the incoming air, so ramjets cannot operate at a standstill and they are most efficient at supersonic speeds. A key trait of ramjet engines is that combustion is done at subsonic speeds. The supersonic incoming air is dramatically slowed through the inlet, where it is then combusted at the much slower, subsonic, speeds. The faster the incoming air is, however, the less efficient it becomes to slow it to subsonic speeds. Therefore, ramjet engines are limited to approximately Mach 5.

Scramjet

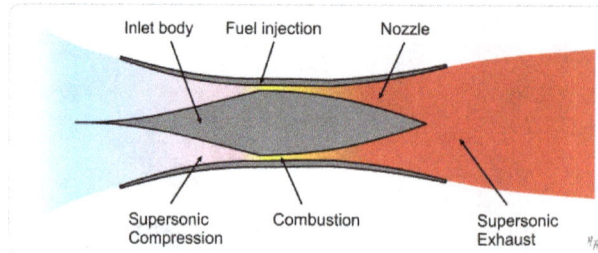

Scramjet engine operation

Scramjets are mechanically very similar to ramjets. Like a ramjet, they consist of an inlet, a combustor, and a nozzle. The primary difference between ramjets and scramjets is that scramjets do not slow the oncoming airflow to subsonic speeds for combustion, they use supersonic combustion instead. The name "scramjet" comes from "Supersonic Combusting Ramjet." Since scramjets use supersonic combustion they can operate at speeds above Mach 6 where traditional ramjets are too inefficient. Another difference between ramjets and scramjets comes from how each type of engine compresses the oncoming airflow: while the inlet provides most of the compression for ramjets, the high speeds at which scramjets operate allow them to take advantage of the compression generated by shock waves, primarily oblique shocks.

Very few scramjet engines have ever been built and flown. In May 2010 the Boeing X-51 set the endurance record for the longest scramjet burn at over 200 seconds.

Non-continuous Combustion

Type	Description	Advantages	Disadvantages
Motorjet	Obsolete type that worked like a turbojet but instead of a turbine driving the compressor a piston engine drives it.	Higher exhaust velocity than a propeller, offering better thrust at high speed	Heavy, inefficient and underpowered. Example: Caproni Campini N.1.
Pulsejet	Air is compressed and combusted intermittently instead of continuously. Some designs use valves.	Very simple design, commonly used on model aircraft	Noisy, inefficient (low compression ratio), works poorly on a large scale, valves on valved designs wear out quickly
Pulse detonation engine	Similar to a pulsejet, but combustion occurs as a detonation instead of a deflagration, may or may not need valves	Maximum theoretical engine efficiency	Extremely noisy, parts subject to extreme mechanical fatigue, hard to start detonation, not practical for current use

Rocket

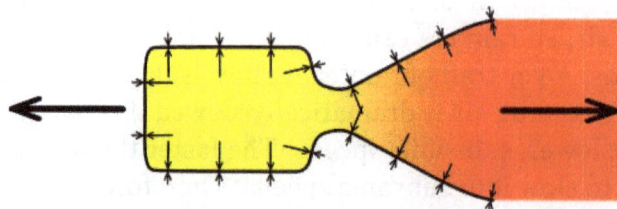

Rocket engine propulsion

The rocket engine uses the same basic physical principles as the jet engine for propulsion via thrust, but is distinct in that it does not require atmospheric air to provide oxygen; the rocket carries all components of the reaction mass. This allows them to operate at arbitrary altitudes and in space.

This type of engine is used for launching satellites, space exploration and manned access, and permitted landing on the moon in 1969.

Rocket engines are used for high altitude flights, or anywhere where very high accelerations are needed since rocket engines themselves have a very high thrust-to-weight ratio.

However, the high exhaust speed and the heavier, oxidizer-rich propellant results in far more propellant use than turbofans. Even so, at extremely high speeds they become energy-efficient.

An approximate equation for the net thrust of a rocket engine is:

$$F_N = \dot{m} g_0 I_{sp-vac} - A_e p$$

Where F_N is the net thrust, $I_{sp(vac)}$ is the specific impulse, g_0 is a standard gravity, $\dot{m}$ is the propellant flow in kg/s, A_e is the cross-sectional area at the exit of the exhaust nozzle, and p is the atmospheric pressure.

Type	Description	Advantages	Disadvantages
Rocket	Carries all propellants and oxidants on board, emits jet for propulsion	Very few moving parts. Mach 0 to Mach 25+; efficient at very high speed (> Mach 5.0 or so). Thrust/weight ratio over 100. No complex air inlet. High compression ratio. Very high-speed (hypersonic) exhaust. Good cost/thrust ratio. Fairly easy to test. Works in a vacuum; indeed, works best outside the atmosphere, which is kinder on vehicle structure at high speed. Fairly small surface area to keep cool, and no turbine in hot exhaust stream. Very high-temperature combustion and high expansion-ratio nozzle gives very high efficiency, at very high speeds.	Needs lots of propellant. Very low specific impulse—typically 100–450 seconds. Extreme thermal stresses of combustion chamber can make reuse harder. Typically requires carrying oxidizer on-board which increases risks. Extraordinarily noisy.

Hybrid

Combined cycle engines simultaneously use 2 or more different jet engine operating principles.

Type	Description	Advantages	Disadvantages
Turborocket	A turbojet where an additional oxidizer such as oxygen is added to the airstream to increase maximum altitude	Very close to existing designs, operates in very high altitude, wide range of altitude and airspeed	Airspeed limited to same range as turbojet engine, carrying oxidizer like LOX can be dangerous. Much heavier than simple rockets.
Air-augmented rocket	Essentially a ramjet where intake air is compressed and burnt with the exhaust from a rocket	Mach 0 to Mach 4.5+ (can also run exoatmospheric), good efficiency at Mach 2 to 4	Similar efficiency to rockets at low speed or exoatmospheric, inlet difficulties, a relatively undeveloped and unexplored type, cooling difficulties, very noisy, thrust/weight ratio is similar to ramjets.

		Easily tested on ground. Very high thrust/weight ratios are possible (~14) together with good fuel efficiency over a wide range of airspeeds, Mach 0-5.5+; this combination of efficiencies may permit launching to orbit, single stage, or very rapid, very long distance intercontinental travel.	
Precooled jets / LACE	Intake air is chilled to very low temperatures at inlet in a heat exchanger before passing through a ramjet and/or turbojet and/or rocket engine.		Exists only at the lab prototyping stage. Examples include RB545, Reaction Engines SABRE, ATREX. Requires liquid hydrogen fuel which has very low density and requires heavily insulated tankage.

Water Jet

A water jet, or pump jet, is a marine propulsion system that utilizes a jet of water. The mechanical arrangement may be a ducted propeller with nozzle, or a centrifugal compressor and nozzle.

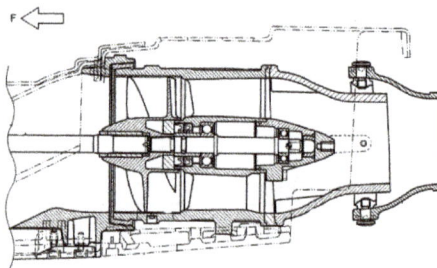

A pump jet schematic.

Type	Description	Advantages	Disadvantages
Water jet	For propelling water rockets and jetboats; squirts water out the back through a nozzle	In boats, can run in shallow water, high acceleration, no risk of engine overload (unlike propellers), less noise and vibration, highly maneuverable at all boat speeds, high speed efficiency, less vulnerable to damage from debris, very reliable, more load flexibility, less harmful to wildlife	Can be less efficient than a propeller at low speed, more expensive, higher weight in boat due to entrained water, will not perform well if boat is heavier than the jet is sized for

General Physical Principles

All jet engines are reaction engines that generate thrust by emitting a jet of fluid rearwards at relatively high speed. The forces on the inside of the engine needed to create this jet give a strong thrust on the engine which pushes the craft forwards.

Jet engines make their jet from propellant from tankage that is attached to the engine (as in a 'rocket') as well as in duct engines (those commonly used on aircraft) by ingesting an external fluid (very typically air) and expelling it at higher speed.

Propelling Nozzle

The propelling nozzle is the key component of all jet engines as it creates the exhaust jet. Propelling nozzles turn internal and pressure energy into high velocity kinetic energy. The total pressure and temperature don't change through the nozzle but their static values drop as the gas speeds up.

The velocity of the air entering the nozzle is low, about Mach 0.4, a prerequisite for minimising pressure losses in the duct leading to the nozzle. The temperature entering the nozzle may be as low as sea level ambient for a fan nozzle in the cold air at cruise altitudes. It may be as high as the 1000K exhaust gas temperature for a supersonic afterburning engine or 2200K with afterburner lit. The pressure entering the nozzle may vary from 1.5 times the pressure outside the nozzle, for a single stage fan, to 30 times for the fastest manned aircraft at mach 3+.

The velocity of the gas leaving a convergent nozzle may be subsonic or sonic (Mach 1) at low flight speeds or supersonic (Mach 3.0 at SR-71 cruise) for a con-di nozzle at higher speeds where the nozzle pressure ratio is increased with the intake ram. The nozzle thrust is highest if the static pressure of the gas reaches the ambient value as it leaves the nozzle. This only happens if the nozzle exit area is the correct value for the nozzle pressure ratio (npr). Since the npr changes with engine thrust setting and flight speed this is seldom the case. Also at supersonic speeds the divergent area is less than required to give complete internal expansion to ambient pressure as a trade-off with external body drag. Whitford gives the F-16 as an example. Other underexpanded examples were the XB-70 and SR-71.

The nozzle size, together with the area of the turbine nozzles, determines the operating pressure of the compressor.

Thrust

Origin of Engine Thrust

The familiar explanation for jet thrust is a "black box" description which only looks at what goes in to the engine, air and fuel, and what comes out, exhaust gas and an unbalanced force. This force, called thrust, is the sum of the momentum difference between entry and exit and any unbalanced pressure force between entry and exit, as explained in "Thrust calculation". As an example, an early turbojet, the Bristol Olympus Mk. 101, had a momentum thrust of 9300 lb. and a pressure thrust of 1800 lb. giving a total of 11,100 lb. Looking inside the "black box" shows that the thrust results from all the unbalanced momentum and pressure forces created within the engine itself. These forces, some forwards and some rearwards, are across all the internal parts, both stationary and rotating, such as ducts, compressors, etc., which are in the primary gas flow which flows through the engine from front to rear. The algebraic sum of all these forces is delivered to the airframe for propulsion. "Flight" gives examples of these internal forces for two early jet engines, the Rolls-Royce Avon Ra.14 and the de Havilland Goblin

Transferring Thrust to the Aircraft

The engine thrust acts along the engine centreline. The aircraft "holds" the engine on the outer casing of the engine at some distance from the engine centreline (at the engine mounts). This arrangement causes the engine casing to bend (known as backbone bending) and the round rotor casings to distort (ovalization). Distortion of the engine structure has to be controlled with suitable mount locations to maintain acceptable rotor and seal clearances and prevent rubbing. A well-publicized example of excessive structural deformation occurred with the original Pratt & Whitney JT9D engine installation in the Boeing 747 aircraft. The engine mounting arrangement had to be revised with the addition of an extra thrust frame to reduce the casing deflections to an acceptable amount.

Rotor Thrust

The rotor thrust on a thrust bearing is not related to the engine thrust. It may even change direction at some RPM. The bearing load is determined by bearing life considerations. Although the aerodynamic loads on the compressor and turbine blades contribute to the rotor thrust they are small compared to cavity loads inside the rotor which result from the secondary air system pressures and sealing diameters on discs, etc. To keep the load within the bearing specification seal diameters are chosen accordingly as, many years ago, on the backface of the impeller in the de Havilland Ghost engine. Sometimes an extra disc known as a balance piston has to be added inside the rotor. An early turbojet example with a balance piston was the Rolls-Royce Avon.

Thrust Calculation

The net thrust (F_N) of a turbojet is given by:

$$F_N = (\dot{m}_{air} + \dot{m}_{fuel})v_e - \dot{m}_{air}v$$

where:

$\dot{m}_{air}$ = the mass rate of air flow through the engine

$\dot{m}_{fuel}$ = the mass rate of fuel flow entering the engine

v_e = the velocity of the jet (the exhaust plume) and is assumed to be less than sonic velocity

v = the velocity of the air intake = the true airspeed of the aircraft

$(\dot{m}_{air} + \dot{m}_{fuel})v_e$ = the nozzle gross thrust (F_G)

$\dot{m}_{air}v$ = the ram drag of the intake air

The above equation applies only for air-breathing jet engines. It does not apply to rocket engines. Most types of jet engine have an air intake, which provides the bulk of the fluid exiting the exhaust. Conventional rocket engines, however, do not have an intake, the oxidizer and fuel both being carried within the vehicle. Therefore, rocket engines do not have ram drag and the gross thrust of the rocket engine nozzle is the net thrust of the engine. Consequently, the thrust characteristics of a rocket motor are different from that of an air breathing jet engine, and thrust is independent of velocity.

If the velocity of the jet from a jet engine is equal to sonic velocity, the jet engine's nozzle is said to be choked. If the nozzle is choked, the pressure at the nozzle exit plane is greater than atmospheric pressure, and extra terms must be added to the above equation to account for the pressure thrust.

The rate of flow of fuel entering the engine is very small compared with the rate of flow of air. If the contribution of fuel to the nozzle gross thrust is ignored, the net thrust is:

$$F_N = \dot{m}_{air}(v_e - v)$$

The velocity of the jet (v_e) must exceed the true airspeed of the aircraft (v) if there is to be a net forward thrust on the aircraft. The velocity (v_e) can be calculated thermodynamically based on adiabatic expansion.

Thrust Augmentation

Thrust augmentation has taken many forms, most commonly to supplement inadequate take-off thrust. Some early jet aircraft needed rocket assistance to take off from high altitude airfields or when the day temperature was high. A more recent aircraft, the Tupolev Tu-22 supersonic bomber, was fitted with four SPRD-63 boosters for take-off. Possibly the most extreme requirement needing rocket assistance, and which was short-lived, was zero-length launching. Almost as extreme, but very common, is catapult assistance from aircraft carriers. Rocket assistance has also been used during flight. The SEPR 841 booster engine was used on the Dassault Mirage for high altitude interception.

Early aft-fan arrangements which added bypass airflow to a turbojet were known as thrust augmentors. The aft-fan fitted to the General Electric CJ805-3 turbojet augmented the take-off thrust from 11,650lb to 16,100lb.

Water, or other coolant, injection into the compressor or combustion chamber and fuel injection into the jetpipe (afterburning/reheat) became standard ways to increase thrust, known as 'wet' thrust to differentiate with the no-augmentation 'dry' thrust.

Coolant injection (pre-compressor cooling) has been used, together with afterburning, to increase thrust at supersonic speeds. The 'Skyburner' McDonnell Douglas F-4 Phantom II set a world speed record using water injection in front of the engine.

At high Mach numbers afterburners supply progressively more of the engine thrust as the thrust from the turbomachine drops off towards zero at which speed the engine pressure ratio (epr) has fallen to 1.0 and all the engine thrust comes from the afterburner. The afterburner also has to make up for the pressure loss across the turbomachine which is a drag item at higher speeds where the epr will be less than 1.0.

Thrust augmentation of existing afterburning engine installations for special short-duration tasks has been the subject of studies for launching small payloads into low earth orbits using aircraft such as McDonnell Douglas F-4 Phantom II, McDonnell Douglas F-15 Eagle, Dassault Rafale and Mikoyan MiG-31, and also for carrying experimental packages to high altitudes using a Lockheed SR-71. In the first case an increase in the existing maximum speed capability is required for orbital launches. In the second case an increase in thrust within the existing speed capability is required. Compressor inlet cooling is used in the first case. A compressor map shows that the airflow reduces with increasing compressor inlet temperature although the compressor is still running at maximum RPM (but reduced aerodynamic speed). Compressor inlet cooling increases the aerodynamic speed and flow and thrust. In the second case a small increase in the maximum mechanical speed and turbine temperature were allowed, together with nitrous oxide injection into the afterburner and simultaneous increase in afterburner fuel flow.

Energy Efficiency Relating to Aircraft Jet Engines

This overview highlights where energy losses occur in complete jet aircraft powerplants or engine installations. It includes mention of inlet and exhaust nozzle losses which become increasingly significant at the high flight speeds achieved by some manned aircraft since only a small proportion, 17% for the SR-71 powerplant and 8% for the Concorde powerplant, of the thrust transmitted to the airframe came from the engine.

A jet engine at rest, as on a test stand, sucks in fuel and tries to thrust itself forward. How well it does this is judged by how much fuel it uses and what force is required to restrain it. This is a measure of its efficiency. If something deteriorates inside the engine (known as performance deterioration) it will be less efficient and this will show when the fuel produces less thrust. If a change is made to an internal part which allows the air/combustion gases to flow more smoothly the engine will be more efficient and use less fuel. A standard definition is used to assess how different things change engine efficiency and also to allow comparisons to be made between different engines. This definition is called specific fuel consumption, or how much fuel is needed to produce one unit of thrust. For example, it will be known for a particular engine design that if some bumps in a bypass duct are smoothed out the air will flow more smoothly giving a pressure loss reduction of x% and y% less fuel will be needed to get the take-off thrust, for example. This understanding comes under the engineering discipline Jet engine performance. How efficiency is affected by forward speed and by supplying energy to aircraft systems is mentioned later.

The efficiency of the engine is controlled primarily by the operating conditions inside the engine which are the pressure produced by the compressor and the temperature of the combustion gases at the first set of rotating turbine blades. The pressure is the highest air pressure in the engine. The turbine rotor temperature is not the highest in the engine but is the highest at which energy transfer takes place (higher temperatures occur in the combustor). The above pressure and temperature are shown on a Thermodynamic cycle diagram.

The efficiency is further modified by how smoothly the air and the combustion gases flow through the engine, how well the flow is aligned (known as incidence angle) with the moving and stationary passages in the compressors and turbines. Non-optimum angles, as well as non-optimum passage and blade shapes can cause thickening and separation of Boundary layers and formation of Shock waves as explained in Effects of Mach number and shock losses in turbomachines. It is important to slow the flow (lower speed means less pressure losses or Pressure drop) when it travels through ducts connecting the different parts. How well the individual components contribute to turning fuel into thrust is quantified by measures like efficiencies for the compressors, turbines and combustor and pressure losses for the ducts. These are shown as lines on a Thermodynamic cycle diagram.

The engine efficiency, or thermal efficiency, known as η_{th}. is dependent on the Thermodynamic cycle parameters, maximum pressure and temperature, and on component efficiencies, $\eta_{compressor}$, $\eta_{combustion}$ and $\eta_{turbine}$ and duct pressure losses.

The engine needs compressed air for itself just to run successfully. This air comes from its own compressor and is called secondary air. It does not contribute to making thrust so makes the engine less efficient. It is used to preserve the mechanical integrity of the engine, to stop parts overheating and to prevent oil escaping from bearings for example. Only some of this air taken from the compressors returns to the turbine flow to contribute to thrust production. Any reduction in the amount needed improves the engine efficiency. Again, it will be known for a particular engine design that a reduced requirement for cooling flow of x% will reduce the specific fuel consumption by y%. In other words, less fuel will be required to give take-off thrust, for example. The engine is more efficient.

All of the above considerations are basic to the engine running on its own and, at the same time, doing nothing useful, i.e. it is not moving an aircraft or supplying energy for the aircraft's electrical, hydraulic and air systems. In the aircraft the engine gives away some of its thrust-producing potential, or fuel, to power these systems. These requirements, which cause installation losses, reduce its efficiency. It is using some fuel that does not contribute to the engine's thrust.

Finally, when the aircraft is flying the propelling jet itself contains wasted kinetic energy after it has left the engine. This is quantified by the term propulsive, or Froude, efficiency η_p and may be reduced by redesigning the engine to give it bypass flow and a lower speed for the propelling jet, for example as a turboprop or turbofan engine. At the same time forward speed increases the η_{th} by increasing the Overall pressure ratio.

The overall efficiency of the engine at flight speed is defined as $\eta_o = \eta_p \eta_{th}$..

The η_o at flight speed depends on how well the intake compresses the air before it is handed over to the engine compressors. The intake compression ratio, which can be as high as 32:1 at Mach 3, adds to that of the engine compressor to give the Overall pressure ratio and η_{th} for the Thermo-dynamic cycle. How well it does this is defined by its pressure recovery or measure of the losses in the intake. Mach 3 manned flight has provided an interesting illustration of how these losses can increase dramatically in an instant. The North American XB-70 Valkyrie and Lockheed SR-71 Blackbird at Mach 3 each had pressure recoveries of about 0.8, due to relatively low losses during the compression process, i.e. through systems of multiple shocks. During an 'unstart' the efficient shock system would be replaced by a very inefficient single shock beyond the inlet and an intake pressure recovery of about 0.3 and a correspondingly low pressure ratio.

The propelling nozzle at speeds above about Mach 2 usually has extra internal thrust losses because the exit area is not big enough as a trade-off with external afterbody drag.

Although a bypass engine improves propulsive efficiency it incurs losses of its own inside the engine itself. Machinery has to be added to transfer energy from the gas generator to a bypass airflow. The low loss from the propelling nozzle of a turbojet is added to with extra losses due to inefficiencies in the added turbine and fan. These may be included in a transmission, or transfer, efficiency η_T. However, these losses are more than made up by the improvement in propulsive efficiency. There are also extra pressure losses in the bypass duct and an extra propelling nozzle.

With the advent of turbofans with their loss-making machinery what goes on inside the engine has been separated by Bennett, for example, between gas generator and transfer machinery giving $\eta_o = \eta_p \eta_{th} \eta_T$..

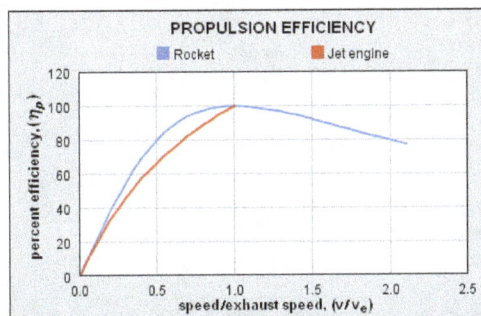

Dependence of propulsion efficiency (η) upon the vehicle speed/exhaust velocity ratio (v/ve) for air-breathing jet and rocket engines.

The energy efficiency (η_o) of jet engines installed in vehicles has two main components:

- *propulsive efficiency* (η_p): how much of the energy of the jet ends up in the vehicle body rather than being carried away as kinetic energy of the jet.

- *cycle efficiency* (η_{th}): how efficiently the engine can accelerate the jet

Even though overall energy efficiency η_o is:

$$\eta_o = \eta_p \eta_{th}$$

for all jet engines the *propulsive efficiency* is highest as the exhaust jet velocity gets closer to the vehicle speed as this gives the smallest residual kinetic energy. For an airbreathing engine an exhaust velocity equal to the vehicle velocity, or a η_p equal to one, gives zero thrust with no net momentum change. The formula for air-breathing engines moving at speed   with an exhaust velocity v_e, and neglecting fuel flow, is:

And for a rocket:

$$\eta_p = \frac{2}{1 + \dfrac{v_e}{v}}$$

$$\eta_p = \frac{2(\dfrac{v}{v_e})}{1 + (\dfrac{v}{v_e})^2}$$

In addition to propulsive efficiency, another factor is *cycle efficiency*; a jet engine is a form of heat engine. Heat engine efficiency is determined by the ratio of temperatures reached in the engine to that exhausted at the nozzle. This has improved constantly over time as new materials have been introduced to allow higher maximum cycle temperatures. For example, composite materials, combining metals with ceramics, have been developed for HP turbine blades, which run at the maximum cycle temperature. The efficiency is also limited by the overall pressure ratio that can be achieved. Cycle efficiency is highest in rocket engines (~60+%), as they can achieve extremely high combustion temperatures. Cycle efficiency in turbojet and similar is nearer to 30%, due to much lower peak cycle temperatures.

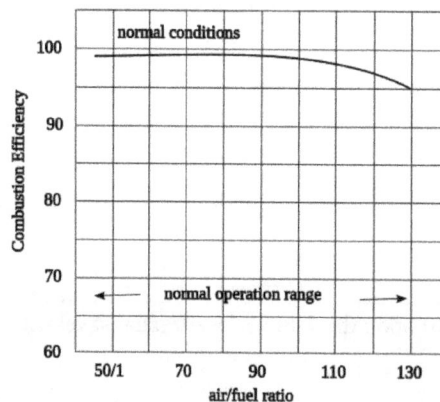

Typical combustion efficiency of an aircraft gas turbine over the operational range.

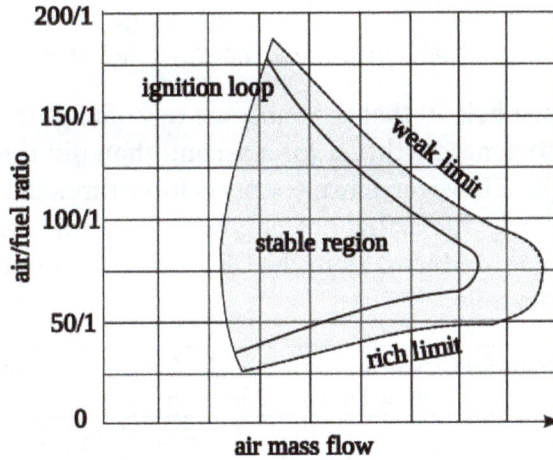

Typical combustion stability limits of an aircraft gas turbine.

The combustion efficiency of most aircraft gas turbine engines at sea level takeoff conditions is almost 100%. It decreases nonlinearly to 98% at altitude cruise conditions. Air-fuel ratio ranges from 50:1 to 130:1. For any type of combustion chamber there is a *rich* and *weak limit* to the air-fuel ratio, beyond which the flame is extinguished. The range of air-fuel ratio between the rich and weak limits is reduced with an increase of air velocity. If the increasing air mass flow reduces the fuel ratio below certain value, flame extinction occurs.

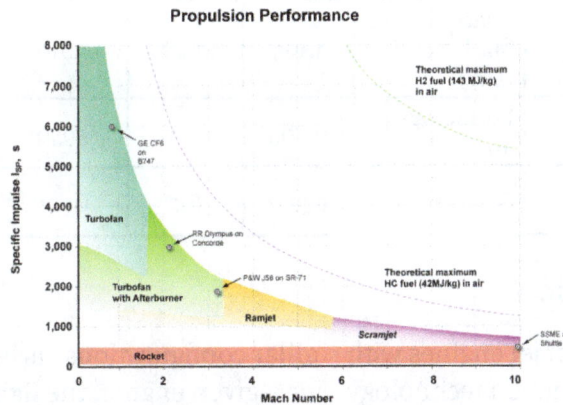

Specific impulse as a function of speed for different jet types with kerosene fuel (hydrogen I_{sp} would be about twice as high). Although efficiency plummets with speed, greater distances are covered. Efficiency per unit distance (per km or mile) is roughly independent of speed for jet engines as a group; however, airframes become inefficient at supersonic speeds.

Consumption of Fuel or Propellant

A closely related (but different) concept to energy efficiency is the rate of consumption of propellant mass. Propellant consumption in jet engines is measured by Specific Fuel Consumption, Specific impulse or Effective exhaust velocity. They all measure the same thing. Specific impulse and effective exhaust velocity are strictly proportional, whereas specific fuel consumption is inversely proportional to the others.

For airbreathing engines such as turbojets, energy efficiency and propellant (fuel) efficiency are much the same thing, since the propellant is a fuel and the source of energy. In rocketry, the propellant is also the exhaust, and this means that a high energy propellant gives better propellant efficiency but can in some cases actually give *lower* energy efficiency.

It can be seen in the table (just below) that the subsonic turbofans such as General Electric's CF6 turbofan use a lot less fuel to generate thrust for a second than did the Concorde's Rolls-Royce/ Snecma Olympus 593 turbojet. However, since energy is force times distance and the distance per second was greater for Concorde, the actual power generated by the engine for the same amount of fuel was higher for Concorde at Mach 2 than the CF6. Thus, the Concorde's engines were more efficient in terms of energy per mile.

Specific fuel consumption (SFC), specific impulse, and effective exhaust velocity numbers for various rocket and jet engines.					
Engine type	Scenario	SFC in lb/ (lbf·h)	SFC in g/ (kN·s)	Specific impulse (s)	Effective exhaust velocity (m/s)
NK-33 rocket engine	Vacuum	10.9	308	331	3250
SSME rocket engine	Space shuttle vacuum	7.95	225	453	4440
Ramjet	Mach 1	4.5	130	800	7800
J-58 turbojet	SR-71 at Mach 3.2 (Wet)	1.9	54	1900	19000
Eurojet EJ200	Reheat	1.7	47	2200	21000
Rolls-Royce/Snecma Olympus 593 turbojet	Concorde Mach 2 cruise (Dry)	1.195	33.8	3010	29500
CF6-80C2B1F turbofan	Boeing 747-400 cruise	0.605	17.1	5950	58400
General Electric CF6 turbofan	Sea level	0.307	8.7	11700	115000

Thrust-to-weight Ratio

The thrust-to-weight ratio of jet engines with similar configurations varies with scale, but is mostly a function of engine construction technology. For a given engine, the lighter the engine, the better the thrust-to-weight is, the less fuel is used to compensate for drag due to the lift needed to carry the engine weight, or to accelerate the mass of the engine.

As can be seen in the following table, rocket engines generally achieve much higher thrust-to-weight ratios than duct engines such as turbojet and turbofan engines. This is primarily because rockets almost universally use dense liquid or solid reaction mass which gives a much smaller volume and hence the pressurisation system that supplies the nozzle is much smaller and lighter for the same performance. Duct engines have to deal with air which is two to three orders of magnitude less dense and this gives pressures over much larger areas, which in turn results in more engineering materials being needed to hold the engine together and for the air compressor.

Jet or rocket engine	Mass (kg)	Mass (lb)	Thrust (kN)	Thrust (lbf)	Thrust-to-weight ratio
RD-0410 nuclear rocket engine	2,000	4,400	35.2	7,900	1.8

Jet or rocket engine	Mass (kg)	Mass (lb)	Thrust (kN)	Thrust (lbf)	Thrust-to-weight ratio
J58 jet engine (SR-71 Blackbird)	2,722	6,001	150	34,000	5.2
Rolls-Royce/Snecma Olympus 593 turbojet with reheat (Concorde)	3,175	7,000	169.2	38,000	5.4
Pratt & Whitney F119	1,800	3,900	91	20,500	7.95
RD-0750 rocket engine, three-propellant mode	4,621	10,188	1,413	318,000	31.2
RD-0146 rocket engine	260	570	98	22,000	38.4
SSME rocket engine (Space Shuttle)	3,177	7,004	2,278	512,000	73.1
RD-180 rocket engine	5,393	11,890	4,152	933,000	78.5
RD-170 rocket engine	9,750	21,500	7,887	1,773,000	82.5
F-1 (Saturn V first stage)	8,391	18,499	7,740.5	1,740,100	94.1
NK-33 rocket engine	1,222	2,694	1,638	368,000	136.7
Merlin 1D rocket engine, full-thrust version	467	1,030	825	185,000	180.1

Rocket thrusts are vacuum thrusts unless otherwise noted

Comparison of Types

Propeller engines handle larger air mass flows, and give them smaller acceleration, than jet engines. Since the increase in air speed is small, at high flight speeds the thrust available to propeller-driven airplanes is small. However, at low speeds, these engines benefit from relatively high propulsive efficiency.

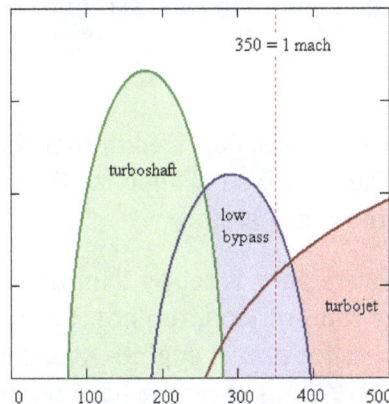

Comparative suitability for (left to right) turboshaft, low bypass and turbojet to fly at 10 km altitude in various speeds. Horizontal axis - speed, m/s. Vertical axis displays engine efficiency.

On the other hand, turbojets accelerate a much smaller mass flow of intake air and burned fuel, but they then reject it at very high speed. When a de Laval nozzle is used to accelerate a hot engine exhaust, the outlet velocity may be locally supersonic. Turbojets are particularly suitable for aircraft travelling at very high speeds.

Turbofans have a mixed exhaust consisting of the bypass air and the hot combustion product gas from the core engine. The amount of air that bypasses the core engine compared to the amount flowing into the engine determines what is called a turbofan's bypass ratio (BPR).

While a turbojet engine uses all of the engine's output to produce thrust in the form of a hot high-velocity exhaust gas jet, a turbofan's cool low-velocity bypass air yields between 30 percent and 70 percent of the total thrust produced by a turbofan system.

The net thrust (F_N) generated by a turbofan is:

$$F_N = \dot{m}_e v_e - \dot{m}_o v_o + BPR(\dot{m}_c v_f)$$

where:

$\dot{m}_e$ = the mass rate of hot combustion exhaust flow from the core engine

$\dot{m}_o$ = the mass rate of total air flow entering the turbofan = $\dot{m}_c + \dot{m}_f$

$\dot{m}_c$ = the mass rate of intake air that flows to the core engine

$\dot{m}_f$ = the mass rate of intake air that bypasses the core engine

v_f = the velocity of the air flow bypassed around the core engine

v_e = the velocity of the hot exhaust gas from the core engine

v_o = the velocity of the total air intake = the true airspeed of the aircraft

BPR = Bypass Ratio

Rocket engines have extremely high exhaust velocity and thus are best suited for high speeds (hypersonic) and great altitudes. At any given throttle, the thrust and efficiency of a rocket motor improves slightly with increasing altitude (because the back-pressure falls thus increasing net thrust at the nozzle exit plane), whereas with a turbojet (or turbofan) the falling density of the air entering the intake (and the hot gases leaving the nozzle) causes the net thrust to decrease with increasing altitude. Rocket engines are more efficient than even scramjets above roughly Mach 15.

Altitude and Speed

With the exception of scramjets, jet engines, deprived of their inlet systems can only accept air at around half the speed of sound. The inlet system's job for transonic and supersonic aircraft is to slow the air and perform some of the compression.

The limit on maximum altitude for engines is set by flammability- at very high altitudes the air becomes too thin to burn, or after compression, too hot. For turbojet engines altitudes of about 40 km appear to be possible, whereas for ramjet engines 55 km may be achievable. Scramjets may theoretically manage 75 km. Rocket engines of course have no upper limit.

At more modest altitudes, flying faster compresses the air at the front of the engine, and this greatly heats the air. The upper limit is usually thought to be about Mach 5-8, as above about Mach 5.5, the atmospheric nitrogen tends to react due to the high temperatures at the inlet and this consumes significant energy. The exception to this is scramjets which may be able to achieve about Mach 15 or more, as they avoid slowing the air, and rockets again have no particular speed limit.

Noise

The noise emitted by a jet engine has many sources. These include, in the case of gas turbine engines, the fan, compressor, combustor, turbine and propelling jet/s.

The propelling jet produces jet noise which is caused by the violent mixing action of the high speed jet with the surrounding air. In the subsonic case the noise is produced by eddies and in the supersonic case by Mach waves. The sound power radiated from a jet varies with the jet velocity raised to the eighth power for velocities up to 2,000 ft/sec and varies with the velocity cubed above 2,000 ft/sec. Thus, the lower speed exhaust jets emitted from engines such as high bypass turbofans are the quietest, whereas the fastest jets, such as rockets, turbojets, and ramjets, are the loudest. For commercial jet aircraft the jet noise has reduced from the turbojet through bypass engines to turbofans as a result of a progressive reduction in propelling jet velocities. For example, the JT8D, a bypass engine, has a jet velocity of 1450 ft/sec whereas the JT9D, a turbofan, has jet velocities of 885 ft/sec (cold) and 1190 ft/sec (hot).

The advent of the turbofan replaced the very distinctive jet noise with another sound known as "buzz saw" noise. The origin is the shockwaves originating at the supersonic fan blades at takeoff thrust.

References

- Ransome-Wallis, Patrick (2001). Illustrated Encyclopedia of World Railway Locomotives. Courier Dover Publications. p. 28. ISBN 0-486-41247-4.

- Zhao, Hua (2010). Advanced Direct Injection Combustion Engine Technologies and Development: Diesel Engines. Woodhead Publishing Limited. p. 8. ISBN 9781845697457.

- The Biodiesel Handbook, Chapter 2—The History of Vegetable Oil Based Diesel Fuels, by Gerhard Knothe, ISBN 978-1-893997-79-0

- Faiz, Asif; Weaver, Christopher S.; Walsh, Michael P. (1996). Air pollution from motor vehicles: Standards and Technologies for Controlling Emissions. World Bank Publications. ISBN 9780821334447.

- Walsh, Philip P.; Paul Fletcher (2004). Gas Turbine Performance (2nd ed.). John Wiley and Sons. p. 25. ISBN 978-0-632-06434-2.

- Hege, John B. (2002). The Wankel Rotary Engine. McFarland. pp. 158–9. ISBN 978-0-7864-1177-1. Retrieved 2012-08-14.

- Faith, Nicholas (1975). Wankel: The Curious Story Behind the Revolutionary Rotary Engine. Stein and Day. p. 219. ISBN 978-0-8128-1719-5.

- Hinckley, Jim; Robinson, Jon G. (2005). The Big Book of Car Culture: The Armchair Guide to Automotive Americana. MBI Publishing. p. 122. ISBN 978-0-7603-1965-9. Retrieved 2011-12-11.

- 'The Wankel Engine', Jan P. Norbye, NSU develops the Wankel, pag 139, and Citroën, pag 305; Chilton, 1971. ISBN 0-8019-5591-2

- Ansdale, Richard F. (1995). Der Wankelmotor. Konstruktion und Wirkungsweise (in German). Motorbuch-Verlag. pp. 73, 91–92, 200. ISBN 978-3-87943-214-1.

- Bensinger, Wolf-Dieter (1973). Rotationskolben-Verbrennungsmotoren (in German). Springer-Verlag. ISBN 978-3-540-05886-1.

- Chinn, Peter (1986). Model Four-Stroke Engines. Wilton, CT USA: Air Age Publishing. pp. 74–81. ISBN 0-911295-04-6.

- Yamaguchi, Jack K (2003). The Mazda RX-8: World's First 4-door, 4-seat Sports Car Plus Complete Histories of Mazda Rotary Engine development and Rotary Racing Around the World. Mazda Motor. ISBN 4-947659-02-5.

- Yamaguchi, Jack K (1985). The New Mazda RX-7 and Mazda Rotary Engine Sports Cars. New York: St. Martin's Press. ISBN 0-312-69456-3.

- "Audi Reveals World's Most Powerful Diesal Passenger Car". Audi UK. 19 September 2006. Archived from the original on 10 February 2007. Retrieved 4 May 2016.

- Ferreira, Omar Campos (March 1998). "Efficiencies of Internal Combustion Engines". Economia & Energia (in Portuguese). Brasil. Retrieved 2016-04-11.

3

Components of Internal Combustion Engine

Cylinder block, piston, combustion chamber, spark plug and camshaft are some of the components of internal combustion engine. Most of these are discussed in this chapter. The topics discussed in the chapter are of great importance to broaden the existing knowledge on the subject matter.

Cylinder Block

A modern straight-six engine block for a passenger car, integrating the crankcase and all cylinders. The cylinder head bolts to the deck surface at top. Many ribs and bosses can be seen on the side of the casting, as well as the passages for cooling fluid opening into the deck.

A cylinder block is an integrated structure comprising the cylinder(s) of a reciprocating engine and often some or all of their associated surrounding structures (coolant passages, intake and exhaust passages and ports, and crankcase). The term engine block is often used synonymously with "cylinder block" (although technically distinctions can be made between en bloc cylinders as a discrete unit versus engine block designs with yet more integration that comprise the crankcase as well).

In the basic terms of machine elements, the various main parts of an engine (such as cylinder(s), cylinder head(s), coolant passages, intake and exhaust passages, and crankcase) are conceptually distinct, and these concepts can all be instantiated as discrete pieces that are bolted together. Such construction was very widespread in the early decades of the commercialization of internal combustion engines (1880s to 1920s), and it is still sometimes used in certain applications where it remains advantageous (especially very large engines, but also some small engines). However, it is no longer the normal way of building most petrol engines and diesel engines, because for any given engine configuration, there are more efficient ways of designing for manufacture (and also for maintenance and repair). These generally involve integrating multiple machine elements into one discrete part, and doing the making (such as casting, stamping, and machining) for multiple elements in one setup with one machine coordinate system (of a machine tool or other piece of manufacturing machinery). This yields lower unit cost of production (and/or maintenance and repair).

A V6 diesel engine block, with both of the cylinder banks as well as the crankcase formed *en bloc*. The large holes are the cylinders, while the small ones are the mounting holes (round) and coolant or oil ducts (oval).

De Dion-Bouton engine with discrete crankcase but with monobloc integration of the cylinders and heads, circa 1905. A discrete crankcase with upper and lower halves (each its own casting) can clearly be seen, with the bottom half constituting both part of the main bearing support and also an oil sump.

Today most engines for cars, trucks, buses, tractors, and so on are built with fairly highly integrated design, so the words "monobloc" and "en bloc" are seldom used in describing them; such construction is often implicit. Thus "engine block", "cylinder block", or simply "block" are the terms likely to be heard in the garage or on the street.

Development Context

The move from extensive use of discrete elements (via separate castings) to extensive integration of elements (such as in most modern engine blocks) was a gradual progression that passed through various phases of monobloc engine development, wherein certain elements were integrated while others remained discrete. This evolution has occurred throughout the history of reciprocating engines, with various instances of every conceptual variation coexisting here and there. The increase in prevalence of ever-more-integrated designs relied on the gradual development of foundry and machining practice for mass production. For example, a practical low-cost V8 engine was not feasible until Ford developed the techniques used to build the Ford flathead V8 engine, which soon also disseminated to the larger society. Today the foundry and machining processes for manufacturing engines are usually highly automated, with a few skilled workers to manage the making of thousands of parts.

Cylinders Integrated Into One or Several Cylinder Blocks

Cylinders are cast in three pairs

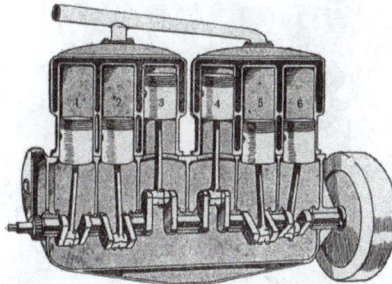

Cylinders are cast in two blocks of three

DB 605 inverted aircraft engine of WW2, with monobloc cylinder blocks and heads

A cylinder block is a unit comprising several cylinders (including their cylinder walls, coolant passages, cylinder sleeves if any, and so forth). In the earliest decades of internal combustion engine development, monobloc cylinder construction was rare; cylinders were usually cast individually. Combining their castings into pairs or triples was an early win of monobloc design.

Each cylinder bank of a V engine (that is, each side of the V) typically comprised one or several cylinder blocks until the 1930s, when mass production methods were developed that allowed the modern form factor of having both banks plus the crankcase entirely integrated.

A wet liner cylinder block features cylinder walls that are entirely removable, which fit into the block by means of special gaskets. They are referred to as "wet liners" because their outer sides come in direct contact with the engine's coolant. In other words, the liner is the entire wall, rather than being merely a sleeve. Wet liner designs are popular with European manufacturers, most notably Renault and Peugeot, who continue to use them to the present. Dry liner designs use either the block's material or a discrete liner inserted into the block to form the backbone of the cylinder wall. Additional sleeves are inserted within, which remain "dry" on their outside, surrounded by the block's material. With either wet or dry liner designs, the liners (or sleeves) can be replaced, potentially allowing overhaul or rebuild without replacement of the block itself; but in reality, they are difficult to remove and install, and for many applications (such as most late-model cars and trucks), an engine will never undergo such a procedure in its working lifespan. It is likelier to be scrapped, with new equipment—engine or entire vehicle—replacing it. This is sometimes rightfully disparaged as a symptom of a throw-away society, but on the other hand, it is actually sometimes more cost-efficient and even environmentally protective to recycle machinery and build new instances with efficient manufacturing processes (and superior machine performance and emission control) than it is to overvalue old machinery and craft production.

Cylinder Blocks and Crankcase Integrated

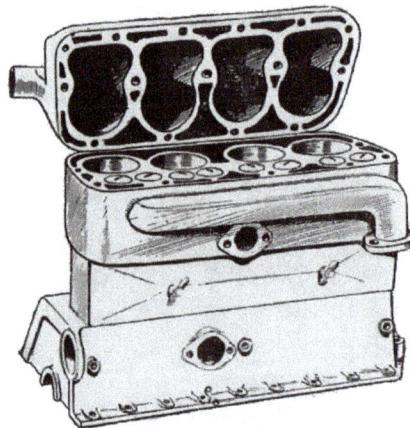

A flathead engine with integral cylinder bank and crankcase. The head is tipped upward to reveal the deck. This example is typical of engines of the 1930s through 1950s.

Casting technology at the dawn of the internal combustion engine could reliably cast either large castings, or castings with complex internal cores to allow for water jackets, but not both simultaneously. Most early engines, particularly those with more than four cylinders, had their cylinders cast as pairs or triplets of cylinders, then bolted to a single crankcase.

As casting techniques improved, an entire cylinder block of 4, 6, or 8 cylinders could be cast as one. This was a simpler construction, thus less expensive (unit-wise) to make. For straight engines, this meant that one engine block could now comprise *all* the cylinders plus the crankcase. Monobloc straight fours, uncommon when the Ford Model T was introduced with one in 1908, became common during the next decade, and monobloc straight sixes followed soon after. By the mid-1920s, both were common, and the straight sixes of General Motors (along with other features that differentiated GM's various makes and models from the Model T) were prying market share away from Ford. (These were all flathead designs.) During that decade, V engines retained a separate block casting for each cylinder bank, with both bolted onto a common crankcase (itself a separate casting). For economy, some engines were designed to use identical castings for each bank, left and right. The complex ducting required for intake and exhaust was too complicated to allow the integration of the banks, except on a few rare engines, such as the Lancia 22½° narrow-angle V12 of 1919, that did manage to use a single block casting for both banks. The hurdles of integrating the banks of the V for common, affordable cars were first overcome by the Ford Motor Company with its Ford flathead V-8, introduced in 1932, which was the first V-8 with a single engine block casting, putting an affordable V-8 into an affordable car for the first time.

The communal water jacket of monobloc designs permitted closer spacing between cylinders. The monobloc design also improved the mechanical stiffness of the engine against bending and the increasingly important torsional twist, as cylinder numbers, engine lengths, and power ratings increased.

Most engines made today, except some unusual V or radial engines, are a monobloc of crankcase and all cylinders. In such cases, the skirts of the cylinder banks form a crankcase area of sorts, which is still often called a crankcase despite no longer being a discrete part.

Engine blocks are normally cast from either a suitable grade of iron or an aluminium alloy. The aluminium block is much lighter in weight, and has better heat transfer to the coolant, but iron blocks retain some advantages and continue to be used by some manufacturers. Because of the use of cylinder liners and bearing shells, the relative softness of aluminium is of no consequence. Some engines have resorted to Plasma transferred wire arc thermal spraying to replace cylinder sleeves and reduce weight.

Combined Block, Head, and Crankcase

Light-duty consumer-grade Honda GC-family small engines use a monobloc design where the cylinder head, block, and half the crankcase share the same casting, termed 'uniblock' by Honda. One reason for this, apart from cost, is to produce an overall lower engine height. Being an air-cooled OHC design, this is possible thanks to current aluminum casting techniques and lack of complex hollow spaces for liquid cooling. The valves are vertical, so as to permit assembly in this confined space. On the other hand, performing basic repairs becomes so time-consuming that the engine can be considered disposable. Commercial-duty Honda GX-family engines (and their many popular knock-offs) have a more conventional design of a single crankcase and cylinder casting, with a separate cylinder head.

Honda produces many other head-block-crankcase monoblocs under a variety of different names, such as the GXV-series. They may all be externally identified by a gasket which would bisect the crankshaft on an approximately 45° angle.

Exhaust valve failure is common and, owing to the monobloc design, so labour-intensive to repair that the engine is normally discarded.

Engine Block, Transmission Case, and Rear Axle Housing as Frame Members

Many farm tractor designs have incorporated their engine block, transmission case, and rear axle housing as frame members. Probably the first was the Fordson tractor, but many others followed. As with many other instances of integration of components into fewer castings, lower unit cost of production was the driver.

An engine block repair shop

Cylinder (Engine)

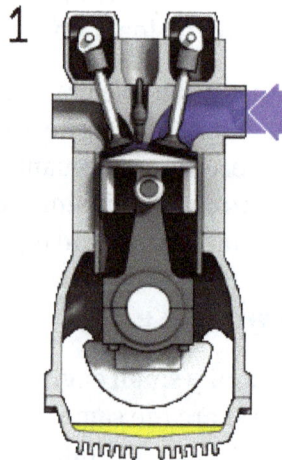

Four-stroke cycle (or Otto cycle)1. Intake2. Compression3. Power4. Exhaust

A cylinder is the central working part of a reciprocating engine or pump, the space in which a piston travels. Multiple cylinders are commonly arranged side by side in a bank, or engine block, which is typically cast from aluminum or cast iron before receiving precision machine work. Cylinders may be sleeved (*lined* with a harder metal) or sleeveless (with a wear-resistant coating such as Nikasil). A sleeveless engine may also be referred to as a "parent-bore engine".

A cylinder's displacement, or swept volume, can be calculated by multiplying its cross-sectional area (the square of half the bore by pi) by the distance the piston travels within the cylinder (the stroke). The engine displacement can be calculated by multiplying the swept volume of one cylinder by the number of cylinders.

Presented symbolically,

$$\text{(Cylinder Volume)} = \pi \cdot \left(\frac{\text{bore}}{2}\right)^2 \cdot \text{Stroke}$$

$$\text{(Engine Displacement)} = \text{(Cylinder Volume)} \cdot \text{(Number of Cylinders)}$$

A piston is seated inside each cylinder by several metal piston rings fitted around its outside surface in machined grooves; typically two for compressional sealing and one to seal the oil. The rings make near contact with the cylinder walls (sleeved or sleeveless), riding on a thin layer of lubricating oil; essential to keep the engine from seizing and necessitating a cylinder wall's durable surface.

During the earliest stage of an engine's life, its initial *breaking-in* or *running-in* period, small irregularities in the metals are encouraged to gradually form congruent grooves by avoiding extreme operating conditions. Later in its life, after mechanical wear has increased the spacing between the piston and the cylinder (with a consequent decrease in power output) the cylinders may be machined to a slightly larger diameter to receive new sleeves (where applicable) and piston rings, a process sometimes known as *reboring*.

Heat Engines

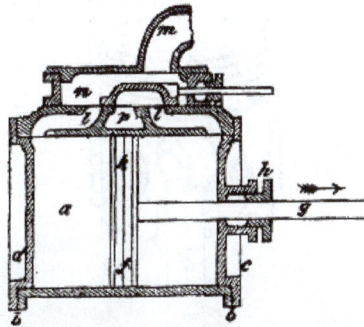

Cylinder with piston in a double acting steam engine

Heat engines, including Stirling engines, are sealed machines using pistons within cylinders to transfer energy from a heat source to a colder reservoir, often using steam or another gas as the working substance. (See Carnot cycle.) The first illustration depicts a longitudinal section of a cylinder in a steam engine. The sliding part at the bottom is the piston, and the upper sliding part is a distribution valve (in this case of the D slide valve type) that directs steam alternately into either end of the cylinder. Refrigerator and air conditioner compressors are heat engines driven in reverse cycle as pumps.

Internal Combustion Engines

Malossi air-cooled cylinder for two-stroke scooters. The exhaust port is visible to the right.

Air-cooled *boxer* engine on a 1954 BMW motorcycle

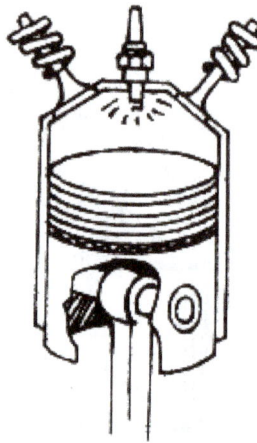

Illustration of an engine cylinder with a cross-section view of the piston, connecting rod, valves and spark plug.

Internal combustion engines operate on the inherent volume change accompanying oxidation of gasoline (petrol), diesel fuel (or some other hydrocarbon) or ethanol, an expansion which is greatly enhanced by the heat produced. They are not classical heat engines since they expel the working substance, which is also the combustion product, into the surroundings.

The reciprocating motion of the pistons is translated into crankshaft rotation via connecting rods. As a piston moves back and forth, a connecting rod changes its angle; its distal end has a rotating link to the crankshaft. A typical four-cylinder automobile engine has a single row of water-cooled cylinders. V engines (V6 or V8) use two angled cylinder banks. The "V" configuration is utilized to create a more compact configuration relative to the number of cylinders. Many other engine configurations exist.

For example, there are also rotary turbines. The Wankel engine is a rotary adaptation of the cylinder-piston concept which has been used by Mazda and NSU in automobiles. Rotary engines are relatively quiet because they lack the clatter of reciprocating motion.

Air-cooled engines generally use individual cases for the cylinders to facilitate cooling. Inline motorcycle engines are an exception, having two-, three-, four-, or even six-cylinder air-cooled units in a common block. Water-cooled engines with only a few cylinders may also use individual cylinder cases, though this makes the cooling system more complex. The Ducati motorcycle company, which for years used air-cooled motors with individual cylinder cases, retained the basic design of their V-twin engine while adapting it to water-cooling.

In some engines, especially French designs, the cylinders have "wet liners". They are formed separately from the main casting so that liquid coolant is free to flow around their outsides. Wet-lined cylinders have better cooling and a more even temperature distribution, but this design makes the engine as a whole somewhat less rigid.

During use, the cylinder is subject to wear from the rubbing action of the piston rings and piston skirt. This is minimized by the thin oil film which coats the cylinder walls and also by a layer of glaze which naturally forms as the engine is run-in, but eventually the cylinder becomes worn and slightly oval in shape, usually necessitating a rebore to an oversize diameter and the fitting of new, oversize pistons. The cylinder does not wear above the highest point reached by the top compression ring of the piston, which can result in a detectable ridge. If an engine is only operated at low rpm for its early life (e.g. in a gently driven automobile) then abruptly used in the higher rpm range (e.g. by a new owner), the slight stretching of the connecting rods at high speed can enable the top compression ring to contact the wear ridge, breaking the ring. For this reason it is important that all engines, once initially run-in, are occasionally "exercised" through their full speed range to develop a tapered wear profile rather than a sharp ridge.

Cylinder Sleeving

Cylinder walls can become very worn or damaged from use. If the engine is not equipped with replaceable sleeves there is a limit to how far the cylinder walls can be bored or worn before the block must be sleeved or replaced. In such cases where the use of a sleeve or liner can restore proper clearances to an engine. Sleeves are made out of iron alloys and are very reliable. A sleeve is installed by a machinist at a machine shop. The engine block is mounted on a precision boring machine where the cylinder is then bored to a size much larger than normal and a new cast-iron sleeve can be inserted with an interference fit. The sleeves can be pressed into place, or they can be held in by a shrink fit. This is done by boring the cylinder (between 3 and 6 thousandths of an inch) smaller than the sleeve being installed, then heating the engine block and while hot, the cold sleeve can be inserted easily. When the engine block cools down it shrink fits around the sleeve holding it into place. Cylinder wall thickness is important to efficient thermal conductivity in the engine. When choosing sleeves, engines have specifications to how thick the cylinder walls should be to prevent overworking the coolant system. Each engine's needs are different, dependent on designed work load duty cycle and energy produced. After selecting and installing the sleeve, the cylinder needs to be finish bored and honed to match the piston. Care needs to be given to the finish of the cylinder walls to prevent improper ring seating at break in.

Seizing

Failed lubrication can cause the pistons or piston rings to seize to the cylinder walls. Seizing can occur during engine use, via overheating and lack of oil, or during storage via condensation and corrosion.

Piston

A piston is a component of reciprocating engines, reciprocating pumps, gas compressors and pneumatic cylinders, among other similar mechanisms. It is the moving component that is contained by a cylinder and is made gas-tight by piston rings. In an engine, its purpose is to transfer force from expanding gas in the cylinder to the crankshaft via a piston rod and/or connecting rod. In a pump, the function is reversed and force is transferred from the crankshaft to the piston for the purpose of compressing or ejecting the fluid in the cylinder. In some engines, the piston also acts as a valve by covering and uncovering ports in the cylinder wall. The petrol enters inside the cylinder and the piston moves upwards and the spark plug produces spark and the petrol is set on fire and it produces an energy that pushes the piston downwards.

Pistons within a sectioned petrol engine

Animation of a piston system.

Piston Engines

Internal Combustion Engines

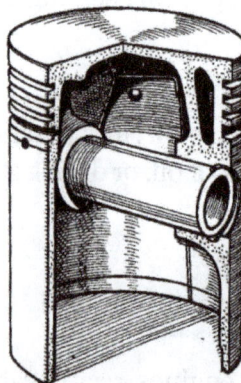

Internal combustion engine piston, sectioned to show the gudgeon pin.

An internal combustion engine is acted upon by the pressure of the expanding combustion gases in the combustion chamber space at the top of the cylinder. This force then acts downwards through the connecting rod and onto the crankshaft. The connecting rod is attached to the piston by a swivelling gudgeon pin (US: wrist pin). This pin is mounted within the piston: unlike the steam engine, there is no piston rod or crosshead (except big two stroke engines).

The pin itself is of hardened steel and is fixed in the piston, but free to move in the connecting rod. A few designs use a 'fully floating' design that is loose in both components. All pins must be prevented from moving sideways and the ends of the pin digging into the cylinder wall, usually by circlips.

Gas sealing is achieved by the use of piston rings. These are a number of narrow iron rings, fitted loosely into grooves in the piston, just below the crown. The rings are split at a point in the rim, allowing them to press against the cylinder with a light spring pressure. Two types of ring are used: the upper rings have solid faces and provide gas sealing; lower rings have narrow edges and a U-shaped profile, to act as oil scrapers. There are many proprietary and detail design features associated with piston rings.

Pistons are cast from aluminium alloys. For better strength and fatigue life, some racing pistons may be forged instead. Early pistons were of cast iron, but there were obvious benefits for engine balancing if a lighter alloy could be used. To produce pistons that could survive engine combustion temperatures, it was necessary to develop new alloys such as Y alloy and Hiduminium, specifically for use as pistons.

A few early gas engines had double-acting cylinders, but otherwise effectively all internal combustion engine pistons are single-acting. During World War II, the US submarine *Pompano* was fitted with a prototype of the infamously unreliable H.O.R. double-acting two-stroke diesel engine. Although compact, for use in a cramped submarine, this design of engine was not repeated.

Trunk Pistons

Trunk pistons are long relative to their diameter. They act both as a piston and cylindrical crosshead. As the connecting rod is angled for much of its rotation, there is also a side force that reacts along the side of the piston against the cylinder wall. A longer piston helps to support this.

Trunk pistons have been a common design of piston since the early days of the reciprocating internal combustion engine. They were used for both petrol and diesel engines, although high speed engines have now adopted the lighter weight slipper piston.

A characteristic of most trunk pistons, particularly for diesel engines, is that they have a groove for an oil ring below the gudgeon pin, in addition to the rings between the gudgeon pin and crown.

The name 'trunk piston' derives from the 'trunk engine', an early design of marine steam engine. To make these more compact, they avoided the steam engine's usual piston rod with separate crosshead and were instead the first engine design to place the gudgeon pin directly within the piston. Otherwise these trunk engine pistons bore little resemblance to the trunk piston; they were extremely large diameter and double-acting. Their 'trunk' was a narrow cylinder mounted in the centre of the piston.

Crosshead Pistons

Large slow-speed Diesel engines may require additional support for the side forces on the piston. These engines typically use crosshead pistons. The main piston has a large piston rod extending downwards from the piston to what is effectively a second smaller-diameter piston. The main piston is responsible for gas sealing and carries the piston rings. The smaller piston is purely a mechanical guide. It runs within a small cylinder as a trunk guide and also carries the gudgeon pin.

Because of the additional weight of these pistons, they are not used for high-speed engines.

Slipper Pistons

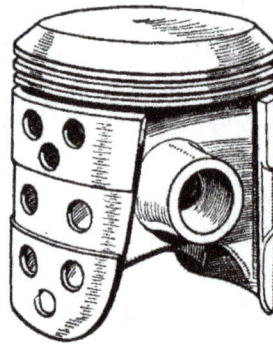

Slipper piston

A slipper piston is a piston for a petrol engine that has been reduced in size and weight as much as possible. In the extreme case, they are reduced to the piston crown, support for the piston rings, and just enough of the piston skirt remaining to leave two lands so as to stop the piston rocking in the bore. The sides of the piston skirt around the gudgeon pin are reduced away from the cylinder wall. The purpose is mostly to reduce the reciprocating mass, thus making it easier to balance the engine and so permit high speeds. A secondary benefit may be some reduction in friction with the cylinder wall, since the area of the skirt, which slides up and down in the cylinder is reduced by half. However most friction is due to the piston rings, which are the parts which actually fit the tightest in the bore and the bearing surfaces of the wrist pin, the benefit is reduced.

Deflector Pistons

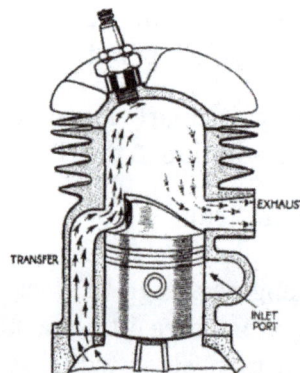

Two-stroke deflector piston

Deflector pistons are used in two-stroke engines with crankcase compression, where the gas flow within the cylinder must be carefully directed in order to provide efficient scavenging. With cross scavenging, the transfer (inlet to the cylinder) and exhaust ports are on directly facing sides of the cylinder wall. To prevent the incoming mixture passing straight across from one port to the other, the piston has a raised rib on its crown. This is intended to deflect the incoming mixture upwards, around the combustion chamber. Much effort, and many different designs of piston crown, went into developing improved scavenging. The crowns developed from a simple rib to a large asymmetric bulge, usually with a steep face on the inlet side and a gentle curve on the exhaust. Despite this, cross scavenging was never as effective as hoped. Most engines today use Schnuerle porting instead. This places a pair of transfer ports in the sides of the cylinder and encourages gas flow to rotate around a vertical axis, rather than a horizontal axis.

Steam Engines

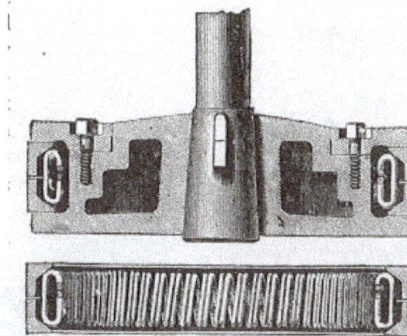

Cast-iron steam engine piston, with a metal piston ring spring-loaded against the cylinder wall.

Steam engines are usually double-acting (i.e. steam pressure acts alternately on each side of the piston) and the admission and release of steam is controlled by slide valves, piston valves or poppet valves. Consequently, steam engine pistons are nearly always comparatively thin discs: their diameter is several times their thickness. (One exception is the trunk engine piston, shaped more like those in a modern internal-combustion engine.) Another factor is that since almost all steam engines use crossheads to translate the force to the drive rod, there are few lateral forces acting to try and "rock" the piston, so a cylinder-shaped piston skirt isn't necessary.

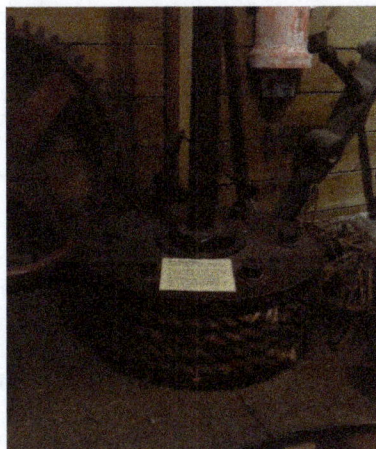

Early (c. 1830) piston for a beam engine. The piston seal is made by turns of wrapped rope.

Pumps

Piston pumps can be used to move liquids or compress gases.

Air Cannons

There are two special type of pistons used in air cannons: close tolerance pistons and double pistons. In close tolerance pistons O-rings serve as a valve, but O-rings are not used in double piston types.

Combustion Chamber

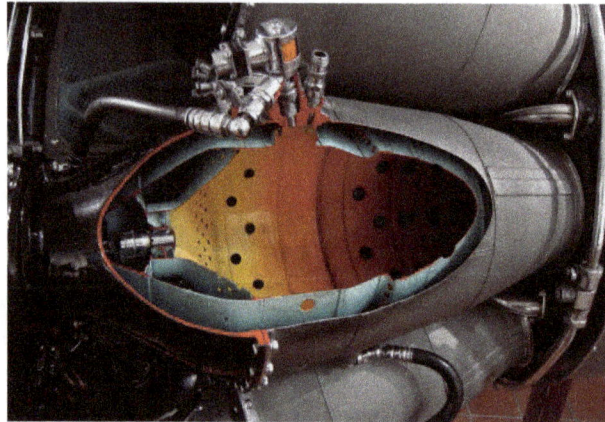

Combustion chamber (named combustor) on a Rolls-Royce Nene turbojet engine

A combustion chamber is that part of an internal combustion engine (ICE) in which the fuel/air mix is burned.

Internal Combustion Engine

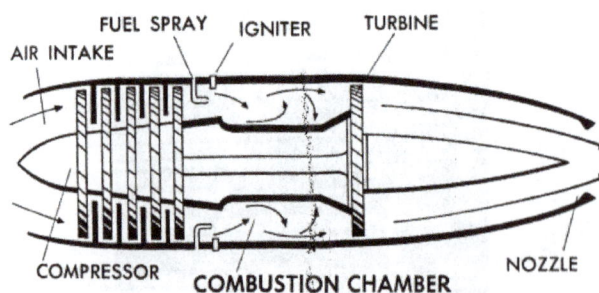

Diagram of jet engine showing the combustion chamber.

ICEs typically comprise reciprocating piston engines, rotary engines, gas turbine and jet turbines.

The combustion process increases the internal energy of a gas, which translates into an increase in temperature, pressure, or volume depending on the configuration. In an enclosure, for example the cylinder of a reciprocating engine, the volume is controlled and the combustion creates an increase in pressure. In a continuous flow system, for example a jet engine combustor, the pressure is controlled and the combustion creates an increase in volume. This increase in pressure or volume can be used to do work, for example, to move a piston on a crankshaft or a turbine disc in a gas turbine. If the gas velocity changes, thrust is produced, such as in the nozzle of a rocket engine.

Petrol (Gasoline) Engine

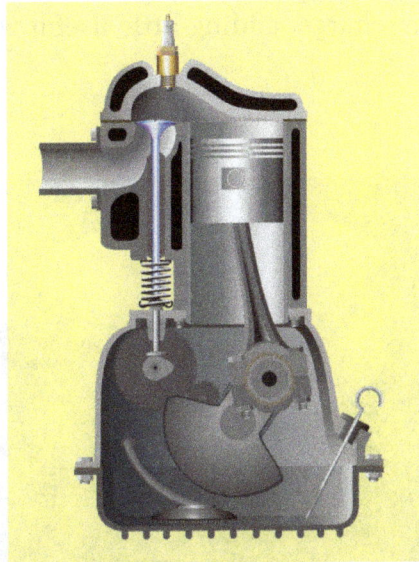

Side-valve engine showing combustion chamber

At top dead centre the pistons of a petrol engine are flush (or nearly flush) with the top of the cylinder block. The combustion chamber may be a recess either in the cylinder head, or in the top of the piston. A design with the combustion chamber in the piston is called a Heron head, where the head is machined flat but the pistons are dished. The Heron head has proved even more thermodynamically efficient than the hemispherical head. Intake valves permit the inflow of a fuel air mix; and exhaust valves allow burnt gases to be scavenged.

Head Types

Various shapes of combustion chamber have been used, such as: L-head (or flathead) for side-valve engines; "bathtub", "hemispherical", and "wedge" for overhead valve engines; and "pent-roof" for engines having 3, 4 or 5 valves per cylinder. The shape of the chamber has a marked effect on power output, efficiency and emissions; the designer's objectives are to burn all of the mixture as completely as possible while avoiding excessive temperatures (which create NOx). This is best achieved with a compact rather than elongated chamber.

Swirl & Squish

The intake valve/port is usually placed to give the mixture a pronounced "swirl" (the term is preferable to "turbulence", which implies movement without overall pattern) above the rising piston, improving mixing and combustion. The shape of the piston top also affects the amount of swirl.

Another design feature to promote turbulence for good fuel/air mixing is "squish", where the fuel/air mix is "squished" at high pressure by the rising piston. Where swirl is particularly important, combustion chambers in the piston may be favoured.

Flame Front

Ignition typically occurs around 15 degrees before top dead centre. The spark plug must be sited so that the flame front can progress throughout the combustion chamber. Good design should avoid narrow crevices where stagnant "end gas" can become trapped, as this gas may detonate violently after the main charge, adding little useful work and potentially damaging the engine.

Diesel Engine

Dished piston for diesel engine

Diesel engines fall into two broad classes:

- Direct injection, where the combustion chamber consists of a dished piston
- Indirect injection, where the combustion chamber is in the cylinder head

Direct injection engines usually give better fuel economy but indirect injection engines can use a lower grade of fuel.

Harry Ricardo was prominent in developing combustion chambers for diesel engines, the best known being the Ricardo Comet.

Gas Turbine

The combustion chamber in gas turbines and jet engines (including ramjets and scramjets) is called the combustor.

The combustor is fed with high pressure air by the compression system, adds fuel and burns the mix and feeds the hot, high pressure exhaust into the turbine components of the engine or out the exhaust nozzle.

Different types of combustors exist, mainly:

- Can type: Can combustors are self-contained cylindrical combustion chambers. Each "can" has its own fuel injector, liner,interconnectors,casing. Each "can" get an air source from individual opening.

- Cannular type: Like the can type combustor, can annular combustors have discrete combustion zones contained in separate liners with their own fuel injectors. Unlike the can combustor, all the combustion zones share a common air casing.

- Annular type: Annular combustors do away with the separate combustion zones and simply have a continuous liner and casing in a ring (the annulus).

Steam Engine

The term combustion chamber is also used to refer to an additional space between the firebox and boiler in a steam locomotive. This space is used to allow further combustion of the fuel, providing greater heat to the boiler.

Large steam locomotives usually have a combustion chamber in the boiler to allow the use of shorter firetubes. This is because:

- Long firetubes have a theoretical advantage in providing a large heating surface but, beyond a certain length, this is subject to diminishing returns.

- Very long firetubes are prone to sagging in the middle.

Micro Combustion Chambers

Micro combustion chambers are the devices in which combustion happens at a very small volume, due to which surface to volume ratio increases which plays a vital role in stabilizing the flame.

Inlet Manifold

Carburetors used as intake runners

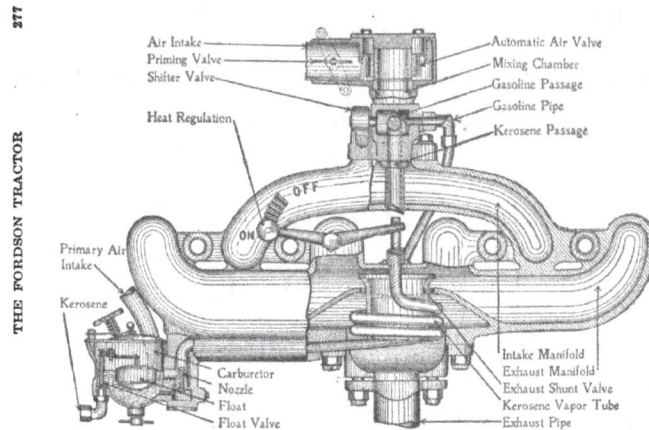

Figure 133.—The Kerosene and Gasoline Vaporizer.

A cutaway view of the intake of the original Fordson tractor
(including the intake manifold, vaporizer, carburetor, and fuel lines).

In automotive engineering, an inlet manifold or intake manifold (in American English) is the part of an engine that supplies the fuel/air mixture to the cylinders. The word *manifold* comes from the Old English word *manigfeald* (from the Anglo-Saxon *manig* [many] and *feald* [repeatedly]) and refers to the multiplying of one (pipe) into many.

In contrast, an exhaust manifold collects the exhaust gases from multiple cylinders into a smaller number of pipes – often down to one pipe.

The primary function of the intake manifold is to *evenly* distribute the combustion mixture (or just air in a direct injection engine) to each intake port in the cylinder head(s). Even distribution is important to optimize the efficiency and performance of the engine. It may also serve as a mount for the carburetor, throttle body, fuel injectors and other components of the engine.

Due to the downward movement of the pistons and the restriction caused by the throttle valve, in a reciprocating spark ignition piston engine, a partial vacuum (lower than atmospheric pressure) exists in the intake manifold. This manifold vacuum can be substantial, and can be used as a source of automobile ancillary power to drive auxiliary systems: power assisted brakes, emission control devices, cruise control, ignition advance, windshield wipers, power windows, ventilation system valves, etc.

This vacuum can also be used to draw any piston blow-by gases from the engine's crankcase. This is known as a positive crankcase ventilation system, in which the gases are burned with the fuel/air mixture.

The intake manifold has historically been manufactured from aluminum or cast iron, but use of composite plastic materials is gaining popularity (e.g. most Chrysler 4-cylinders, Ford Zetec 2.0, Duratec 2.0 and 2.3, and GM's Ecotec series).

Turbulence

The carburetor or the fuel injectors spray fuel droplets into the air in the manifold. Due to electrostatic forces some of the fuel will form into pools along the walls of the manifold, or may converge into larger droplets in the air. Both actions are undesirable because they create inconsistencies in the air-fuel ratio. Turbulence in the intake causes forces of uneven proportions in varying vectors to be applied to the fuel, aiding in atomization. Better atomization allows for a more complete burn of all the fuel and helps reduce engine knock by enlarging the flame front. To achieve this turbulence it is a common practice to leave the surfaces of the intake and intake ports in the cylinder head rough and unpolished.

Only a certain degree of turbulence is useful in the intake. Once the fuel is sufficiently atomized additional turbulence causes unneeded pressure drops and a drop in engine performance.

Volumetric Efficiency

Comparison of a stock intake manifold for a Volkswagen 1.8T engine (top) to a custom-built one used in competition (bottom). In the custom-built manifold, the runners to the intake ports on the cylinder head are much wider and more gently tapered. This difference improves the volumetric efficiency of the engine's fuel/air intake.

The design and orientation of the intake manifold is a major factor in the volumetric efficiency of an engine. Abrupt contour changes provoke pressure drops, resulting in less air (and/or fuel) entering the combustion chamber; high-performance manifolds have smooth contours and gradual transitions between adjacent segments.

Modern intake manifolds usually employ *runners*, individual tubes extending to each intake port on the cylinder head which emanate from a central volume or "plenum" beneath the carburetor. The purpose of the runner is to take advantage of the Helmholtz resonance property of air. Air flows at considerable speed through the open valve. When the valve closes, the air that has not yet entered the valve still has a lot of momentum and compresses against the valve, creating a pocket of high pressure. This high-pressure air begins to equalize with lower-pressure air in the manifold. Due to the air's inertia, the equalization will tend to oscillate: At first the air in the runner will be at a lower pressure than the manifold. The air in the manifold then tries to equalize back into the runner, and the oscillation repeats. This process occurs at the speed of sound, and in most manifolds travels up and down the runner many times before the valve opens again.

The smaller the cross-sectional area of the runner, the higher the pressure changes on resonance for a given airflow. This aspect of Helmholtz resonance reproduces one result of the Venturi effect. When the piston accelerates downwards, the pressure at the output of the intake runner is reduced.

This low pressure pulse runs to the input end, where it is converted into an over-pressure pulse. This pulse travels back through the runner and rams air through the valve. The valve then closes.

To harness the full power of the Helmholtz resonance effect, the opening of the intake valve must be timed correctly, otherwise the pulse could have a negative effect. This poses a very difficult problem for engines, since valve timing is dynamic and based on engine speed, whereas the pulse timing is static and dependent on the length of the intake runner and the speed of sound. The traditional solution has been to tune the length of the intake runner for a specific engine speed where maximum performance is desired. However, modern technology has given rise to a number of solutions involving electronically controlled valve timing (for example Valvetronic), and dynamic intake geometry (see below).

As a result of "resonance tuning", some naturally aspirated intake systems operate at a volumetric efficiency above 100%: the air pressure in the combustion chamber before the compression stroke is greater than the atmospheric pressure. In combination with this intake manifold design feature, the exhaust manifold design, as well as the exhaust valve opening time can be so calibrated as to achieve greater evacuation of the cylinder. The exhaust manifolds achieve a vacuum in the cylinder just before the piston reaches top dead center. The opening inlet valve can then—at typical compression ratios—fill 10% of the cylinder before beginning downward travel. Instead of achieving higher pressure in the cylinder, the inlet valve can stay open after the piston reaches bottom dead center while the air still flows in.

In some engines the intake runners are straight for minimal resistance. In most engines, however, the runners have curves...and some very convoluted to achieve desired runner length. These turns allow for a more compact manifold, with denser packaging of the whole engine, as a result. Also, these "snaked" runners are needed for some variable length/ split runner designs, and allow the size of the plenum to be reduced. In an engine with at least six cylinders the averaged intake flow is nearly constant and the plenum volume can be smaller. To avoid standing waves within the plenum it is made as compact as possible. The intake runners each use a smaller part of the plenum surface than the inlet, which supplies air to the plenum, for aerodynamic reasons. Each runner is placed to have nearly the same distance to the main inlet. Runners whose cylinders fire close after each other, are not placed as neighbors.

"180-degree intake manifolds"....Originally designed for carburetor V8 engines, the two plane, split plenum intake manifold separates the intake pulses which the manifold experiences by 180 degrees in the firing order. This minimizes interference of one cylinder's pressure waves with those of another, giving better torque from smooth mid-range flow. Such manifolds may have been originally designed for either two- or four-barrel carburetors, but now are used with both throttle-body and multi-point fuel injection. An example of the latter is the Honda J engine which converts to a single plane manifold around 3500 rpm for greater peak flow and horsepower.

"Heat Riser"....now obsolete, earlier manifolds ...with 'wet runners' for carbureted engines...used exhaust gas diversion through the intake manifold to provide vaporizing heat. The amount of exhaust gas flow diversion was controlled by a heat riser valve in the exhaust manifold, and employed a bi-metallic spring which changed tension according to the heat in the manifold. Today's fuel-injected engines do not require such devices.

Variable-length Intake Manifold

Lower intake manifold on a 1999 Mazda Miata engine, showing components of a variable length intake system.

Variable-Length Intake Manifold (VLIM) is an internal combustion engine manifold technology. Four common implementations exist. First, two discrete intake runners with different length are employed, and a butterfly valve can close the short path. Second the intake runners can be bent around a common plenum, and a sliding valve separates them from the plenum with a variable length. Straight high-speed runners can receive plugs, which contain small long runner extensions. The plenum of a 6- or 8-cylinder engine can be parted into halves, with the even firing cylinders in one half and the odd firing cylinders in the other part. Both sub-plenums and the air intake are connected to an Y (sort of main plenum). The air oscillates between both sub-plenums, with a large pressure oscillation there, but a constant pressure at the main plenum. Each runner from a sub plenum to the main plenum can be changed in length. For V engines this can be implemented by parting a single large plenum at high engine speed by means of sliding valves into it when speed is reduced.

As the name implies, VLIM can vary the length of the intake tract in order to optimize power and torque, as well as provide better fuel efficiency.

There are two main effects of variable intake geometry:

- Venturi effect - At low rpm, the speed of the airflow is increased by directing the air through a path with limited capacity (cross-sectional area). The larger path opens when the load increases so that a greater amount of air can enter the chamber. In dual overhead cam (DOHC) designs, the air paths are often connected to separate intake valves so the shorter path can be excluded by deactivating the intake valve itself.

- Pressurization - A tuned intake path can have a light pressurizing effect similar to a low-pressure supercharger due to Helmholtz resonance. However, this effect occurs only over a narrow engine speed range which is directly influenced by intake length. A variable intake can create two or more pressurized "hot spots." When the intake air speed is higher, the dynamic pressure pushing the air (and/or mixture) inside the engine is increased. The dynamic pressure is proportional to the square of the inlet air speed, so by making the passage narrower or longer the speed/dynamic pressure is increased.

Many automobile manufacturers use similar technology with different names. Another common term for this technology is Variable Resonance Induction System (VRIS).

Spark Plug

Spark plug with single side electrode

A spark plug (sometimes, in British English, a sparking plug, and, colloquially, a plug) is a device for delivering electric current from an ignition system to the combustion chamber of a spark-ignition engine to ignite the compressed fuel/air mixture by an electric spark, while containing combustion pressure within the engine. A spark plug has a metal threaded shell, electrically isolated from a central electrode by a porcelain insulator. The central electrode, which may contain a resistor, is connected by a heavily insulated wire to the output terminal of an ignition coil or magneto. The spark plug's metal shell is screwed into the engine's cylinder head and thus electrically grounded. The central electrode protrudes through the porcelain insulator into the combustion chamber, forming one or more spark gaps between the inner end of the central electrode and usually one or more protuberances or structures attached to the inner end of the threaded shell and designated the *side*, *earth*, or *ground* electrode(s).

Spark plugs may also be used for other purposes; in Saab Direct Ignition when they are not firing, spark plugs are used to measure ionization in the cylinders – this ionic current measurement is used to replace the ordinary cam phase sensor, knock sensor and misfire measurement function. Spark plugs may also be used in other applications such as furnaces wherein a combustible fuel/air mixture must be ignited. In this case, they are sometimes referred to as flame igniters.

History

In 1860 Étienne Lenoir used an electric spark plug in his gas engine, the first internal combustion piston engine and is generally credited with the invention of the spark plug.

Early patents for spark plugs included those by Nikola Tesla (in U.S. Patent 609,250 for an ignition timing system, 1898), Frederick Richard Simms (GB 24859/1898, 1898) and Robert Bosch (GB 26907/1898). But only the invention of the first commercially viable high-voltage spark plug as part of a magneto-based ignition system by Robert Bosch's engineer Gottlob Honold in 1902 made possible the development of the spark-ignition engine. Subsequent manufacturing improvements can also be credited to Albert Champion, the Lodge brothers, sons of Sir Oliver Lodge, who developed and manufactured their father's idea and also Kenelm Lee Guinness, of the Guinness brewing family, who developed the KLG brand. Helen Blair Bartlett also played a vital role in making the insulator in 1930.

Operation

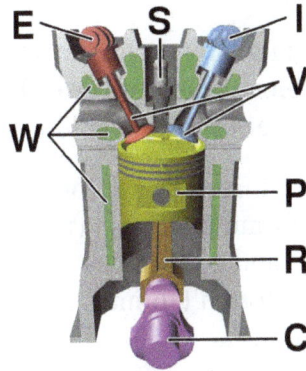

Components of a typical, four stroke cycle, DOHC piston engine.

- (E) Exhaust camshaft

- (I) Intake camshaft

- (S) Spark plug

- (V) Valves

- (P) Piston

- (R) Connecting rod

- (C) Crankshaft

- (W) Water jacket for coolant flow

The plug is connected to the high voltage generated by an ignition coil or magneto. As current flows from the coil, a voltage develops between the central and side electrodes. Initially no current can flow because the fuel and air in the gap is an insulator, but as the voltage rises further it begins to change the structure of the gases between the electrodes. Once the voltage exceeds the dielectric strength of the gases, the gases become ionized. The ionized gas becomes a conductor and allows current to flow across the gap. Spark plugs usually require voltage of 12,000–25,000 volts or more to "fire" properly, although it can go up to 45,000 volts. They supply higher current during the discharge process, resulting in a hotter and longer-duration spark.

As the current of electrons surges across the gap, it raises the temperature of the spark channel to 60,000 K. The intense heat in the spark channel causes the ionized gas to expand very quickly, like a small explosion. This is the "click" heard when observing a spark, similar to lightning and thunder.

The heat and pressure force the gases to react with each other, and at the end of the spark event there should be a small ball of fire in the spark gap as the gases burn on their own. The size of this fireball, or kernel, depends on the exact composition of the mixture between the electrodes and the level of combustion chamber turbulence at the time of the spark. A small kernel will make the engine run as though the ignition timing was retarded, and a large one as though the timing was advanced.

Spark Plug Construction

A spark plug is composed of a shell, insulator and the central conductor. It passes through the wall of the combustion chamber and therefore must also seal the combustion chamber against high pressures and temperatures without deteriorating over long periods of time and extended use.

Spark plugs are specified by size, either thread or nut (often referred to as *Euro*), sealing type (taper or crush washer), and spark gap. Common thread (nut) sizes in Europe are 10 mm (16 mm), 14 mm (21 mm; sometimes, 16 mm), and 18 mm (24 mm, sometimes, 21 mm). In the United States, common thread (nut) sizes are 10mm (16mm), 12mm (14mm, 16mm or 17.5mm), 14mm (16mm, 20.63mm) and 18mm (20.63mm).

Parts of the Plug

Terminal

The top of the spark plug contains a terminal to connect to the ignition system. The exact terminal construction varies depending on the use of the spark plug. Most passenger car spark plug wires snap onto the terminal of the plug, but some wires have eyelet connectors which are fastened onto the plug under a nut. Plugs which are used for these applications often have the end of the terminal serve a double purpose as the nut on a thin threaded shaft so that they can be used for either type of connection.

Insulator

The main part of the insulator is typically made from sintered alumina, a very hard ceramic material with high dielectric strength, printed with the manufacturer's name and identifying marks, then glazed to improve resistance to surface spark tracking. Its major functions are to provide mechanical support and electrical insulation for the central electrode, while also providing an extended spark path for flashover protection. This extended portion, particularly in engines with deeply recessed plugs, helps extend the terminal above the cylinder head so as to make it more readily accessible.

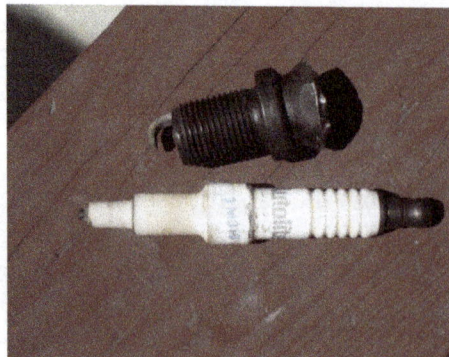

Dissected modern spark plug showing the one-piece sintered alumina insulator. The lower portion is unglazed.

A further feature of sintered alumina is its good heat conduction – reducing the tendency for the insulator to glow with heat and so light the mixture prematurely.

Ribs

By lengthening the surface between the high voltage terminal and the grounded metal case of the spark plug, the physical shape of the ribs functions to improve the electrical insulation and prevent electrical energy from leaking along the insulator surface from the terminal to the metal case. The disrupted and longer path makes the electricity encounter more resistance along the surface of the spark plug even in the presence of dirt and moisture. Some spark plugs are manufactured without ribs; improvements in the dielectric strength of the insulator make them less important.

Insulator Tip

On modern (post 1930s) spark plugs, the tip of the insulator protruding into the combustion chamber is the same sintered aluminium oxide (alumina) ceramic as the upper portion, merely unglazed. It is designed to withstand 650 °C (1,200 °F) and 60 kV.

The dimensions of the insulator and the metal conductor core determine the heat range of the plug. Short insulators are usually "cooler" plugs, while "hotter" plugs are made with a lengthened path to the metal body, though this also depends on the thermally conductive metal core.

Two spark plugs in comparison views in multiple angles, one of which is consumed regularly, while the other has the insulating ceramic broken and the central electrode shortened, due to manufacturing defects and / or temperature swing

Older spark plugs, particularly in aircraft, used an insulator made of stacked layers of mica, compressed by tension in the centre electrode.

With the development of leaded petrol in the 1930s, lead deposits on the mica became a problem and reduced the interval between needing to clean the spark plug. Sintered alumina was developed by Siemens in Germany to counteract this. Sintered alumina is a superior material to mica or porcelain because it is a relatively good thermal conductor for a ceramic, it maintains good mechanical strength and (thermal) shock resistance at higher temperatures, and this ability to run hot allows it to be run at "self cleaning" temperatures without rapid degradation. It also allows a simple single piece construction at low cost but high mechanical reliability.

Seals

Because the spark plug also seals the combustion chamber or the engine when installed, seals are required to ensure there is no leakage from the combustion chamber. The internal seals of modern plugs are made of compressed glass/metal powder, but old style seals were typically made by the use of a multi-layer braze. The external seal is usually a crush washer, but some manufacturers use the cheaper method of a taper interface and simple compression to attempt sealing.

Metal Case/Shell

The metal case/shell (or the *jacket*, as many people call it) of the spark plug withstands the torque of tightening the plug, serves to remove heat from the insulator and pass it on to the cylinder head, and acts as the ground for the sparks passing through the central electrode to the side electrode. Spark plug threads are cold rolled to prevent thermal cycle fatigue. It's important to install spark plugs with the correct "reach," or thread length. Spark plugs can vary in reach from .0375" to 1.043", such for automotive and small engine applications. Also, a marine spark plug's shell is double-dipped, zinc-chromate coated metal.

Central Electrode

Central and lateral electrodes

The central electrode is connected to the terminal through an internal wire and commonly a ceramic series resistance to reduce emission of RF noise from the sparking. Non-resistor spark plugs, commonly sold without an "R" in the plug type part number, lack this element to reduce electro-magnetic interference with radios and other sensitive equipment. The tip can be made of a combination of copper, nickel-iron, chromium, or noble metals. In the late 1970s, the development of engines reached a stage where the heat range of conventional spark plugs with solid nickel alloy centre electrodes was unable to cope with their demands. A plug that was cold enough to cope with the demands of high speed driving would not be able to burn off the carbon deposits caused by stop–start urban conditions, and would foul in these conditions, making the engine misfire. Similarly, a plug that was hot enough to run smoothly in town could melt when called upon to cope with extended high speed running on motorways. The answer to this problem, devised by the spark plug manufacturers, was to use a different material and design for the centre electrode that would be able to carry the heat of combustion away from the tip more effectively than a solid nickel alloy could. Copper was the material chosen for the task and a method for manufacturing the copper-cored centre electrode was created by Floform.

The central electrode is usually the one designed to eject the electrons (the cathode, i.e. negative polarity) because it is the hottest (normally) part of the plug; it is easier to emit electrons from a hot surface, because of the same physical laws that increase emissions of vapor from hot surfaces (see thermionic emission). In addition, electrons are emitted where the electrical field strength is greatest; this is from wherever the radius of curvature of the surface is smallest, from a sharp point or edge rather than a flat surface (see corona discharge). It would be easiest to pull electrons from a pointed electrode but a pointed electrode would erode after only a few seconds. Instead, the electrons emit from the sharp edges of the end of the electrode; as these edges erode, the spark becomes weaker and less reliable.

At one time it was common to remove the spark plugs, clean deposits off the ends either manually or with specialized sandblasting equipment and file the end of the electrode to restore the sharp edges, but this practice has become less frequent for two reasons:

1. cleaning with tools such as a wire brush leaves traces of metal on the insulator which can provide a weak conduction path and thus weaken the spark (increasing emissions)

2. plugs are so cheap relative to labor cost, economics dictate replacement, particularly with modern long-life plugs.

The development of noble metal high temperature electrodes (using metals such as yttrium, iridium, tungsten, or palladium, as well as the relatively high value platinum, silver or gold) allows the use of a smaller center wire, which has sharper edges but will not melt or corrode away. These materials are used because of their high melting points and durability, not because of their electrical conductivity (which is irrelevant in series with the plug resistor or wires). The smaller electrode also absorbs less heat from the spark and initial flame energy. At one point, Firestone marketed plugs with polonium in the tip, under the (questionable) theory that the radioactivity would ionize the air in the gap, easing spark formation.

Side (Ground, Earth) Electrode

The side electrode (also known as the "ground strap") is made from high nickel steel and is welded or hot forged to the side of the metal shell. The side electrode also runs very hot, especially on projected nose plugs. Some designs have provided a copper core to this electrode, so as to increase heat conduction. Multiple side electrodes may also be used, so that they don't overlap the central electrode.

Since electrons flow in the opposite direction (from "-" to "+" electrode) than electrical current itself (from "+" to "-" electrode), it is actually more often from the surface of the side electrode, that is grounded to "-" electrode of the car battery and alternator, where electrons are ejected from towards central electrode. That's why it is important to have edge shaped area or even more areas there (just like some Bosch spark plugs with two edges in spear-like head), that are least exposed to the erosion forces from combustion process thus keeping ignition system in good working condition as long as possible (10.000 – 25.000 km for common spark plugs). Edge or pike is the area from where the spark jumps off most easily (if there are different options for current with similar gap between "+" and "-" electrode) and with less voltage drop (since the gap creates resistance in the circuit). Therefore, some spark plugs can be easily and sometimes even repeatedly renewed by just mild re-sharping these edges (as long as other vital parts and characteristics of spark plugs are not compromised) to the original working state or even re-modelating the original side electrode design (which is very tricky if you are not qualified electrician or physicist). But one has to be careful with side electrode welded joint while repeated re-adjusting gap between electrodes because of metal fatigue, since loosen part of the side electrode in cylinder chamber will result immediate ignition malfunction and almost certainly result in serious damage of special surface-treated cylinder walls or piston sealing rings. The very same problem can cause protruding flakes from poor quality of welding, since they are prone to loosen during prolonged service life of common spark plug. Therefore, present flakes should be treated by rubbing off as well, if spark plug is renewed and used over recommended service life interval.

Spark Plug Gap

Gap gauge: A disk with sloping edge; the edge is thicker going counter-clockwise, and a spark plug will be hooked along the edge to check the gap.

Spark plugs are typically designed to have a spark gap which can be adjusted by the technician installing the spark plug, by bending the ground electrode slightly. The same plug may be specified for several different engines, requiring a different gap for each. Spark plugs in automobiles generally have a gap between 0.6–1.8 mm (0.024"–0.070"). The gap may require adjustment from the out-of-the-box gap.

A *spark plug gap gauge* is a disc with a sloping edge, or with round wires of precise diameters, and is used to measure the gap. Use of a feeler gauge with flat blades instead of round wires, as is used on distributor points or valve lash, will give erroneous results, due to the shape of spark plug electrodes. The simplest gauges are a collection of keys of various thicknesses which match the desired gaps and the gap is adjusted until the key fits snugly. With current engine technology, universally incorporating solid state ignition systems and computerized fuel injection, the gaps used are larger on average than in the era of carburetors and breaker point distributors, to the extent that spark plug gauges from that era cannot always measure the required gaps of current cars. Vehicles using compressed natural gas generally require narrower gaps than vehicles using gasoline.

The gap adjustment can be crucial to proper engine operation. A narrow gap may give too small and weak a spark to effectively ignite the fuel-air mixture, but the plug will almost always fire on each cycle. A gap that is too wide might prevent a spark from firing at all or may misfire at high speeds, but will usually have a spark that is strong for a clean burn. A spark which intermittently fails to ignite the fuel-air mixture may not be noticeable directly, but will show up as a reduction in the engine's power and fuel efficiency.

Variations on the Basic Design

Over the years variations on the basic spark plug design have attempted to provide either better ignition, longer life, or both. Such variations include the use of two, three, or four equally spaced ground electrodes surrounding the central electrode. Other variations include using a recessed central electrode surrounded by the spark plug thread, which effectively becomes the ground electrode (see "surface-discharge spark plug", below). Also there is the use of a V-shaped notch in the tip of the ground electrode. Multiple ground electrodes generally provide longer life, as when the spark gap widens due to electric discharge wear, the spark moves to another closer ground electrode. The disadvantage of multiple ground electrodes is that a shielding effect can occur in the engine combustion chamber inhibiting the flame face as the fuel air mixture burns. This can result in a less efficient burn and increased fuel consumption.

Spark plug with two side (ground) electrodes

Surface-discharge Spark Plug

A piston engine has a part of the combustion chamber that is always out of reach of the piston; and this zone is where the conventional spark plug is located. A Wankel engine has a permanently varying combustion area; and the spark plug is inevitably swept by the tip seals. Clearly, if a spark plug were to protrude into the Wankel's combustion chamber it would foul the rotating tip; and if the plug were recessed to avoid this, the sunken spark might lead to poor combustion. So a new type of "surface discharge" plug was developed for the Wankel. Such a plug presents an almost flat face to the combustion chamber. A stubby centre electrode projects only very slightly; and the entire earthed body of the plug acts as the side electrode. The advantage is that the plug sits just beneath the tip-seal that sweeps over it, keeping the spark accessible to the fuel/air mixture. The "plug gap" remains constant throughout its life; and the spark path will continually vary (instead of darting from the centre to the side electrode as in a conventional plug). Whereas a conventional side electrode will (admittedly, rarely) come adrift in use and potentially cause engine damage, this is impossible with a surface discharge plug, as there is nothing to break off. Surface-discharge spark plugs have been produced by inter alia, Denso, NGK, Champion and Bosch.

Sealing to the Cylinder Head

Old spark plug removed from a car, new one ready to install.

Most spark plugs seal to the cylinder head with a single-use hollow or folded metal washer which is crushed slightly between the flat surface of the head and that of the plug, just above the threads. Some spark plugs have a tapered seat that uses no washer. The torque for installing these plugs is supposed to be lower than a washer-sealed plug. Spark plugs with tapered seats should never be installed in vehicles with heads requiring washers, and vice versa. Otherwise, a poor seal or incorrect reach would result because of the threads not properly seating in the heads.

Tip Protrusion

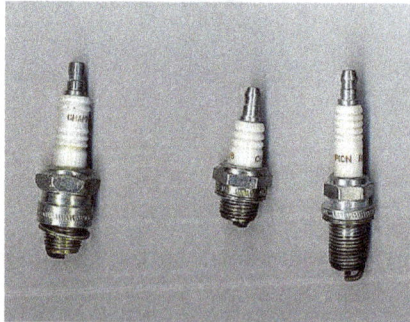

Different spark plug sizes. The left and right plug are identical in threading, electrodes, tip protrusion, and heat range. The centre plug is a compact variant, with smaller hex and porcelain portions outside the head, to be used where space is limited. The rightmost plug has a longer threaded portion, to be used in a thicker cylinder head.

The length of the threaded portion of the plug should be closely matched to the thickness of the head. If a plug extends too far into the combustion chamber, it may be struck by the piston, damaging the engine internally. Less dramatically, if the threads of the plug extend into the combustion chamber, the sharp edges of the threads act as point sources of heat which may cause pre-ignition; in addition, deposits which form between the exposed threads may make it difficult to remove the plugs, even damaging the threads on aluminium heads in the process of removal. The protrusion of the tip into the chamber also affects plug performance, however; the more centrally located the spark gap is, generally the better the ignition of the air-fuel mixture will be, although experts believe the process is more complex and dependent on combustion chamber shape. On the other hand, if an engine is "burning oil", the excess oil leaking into the combustion chamber tends to foul the plug tip and inhibit the spark; in such cases, a plug with less protrusion than the engine would normally call for often collects less fouling and performs better, for a longer period. In fact, special "anti-fouling" adapters are sold which fit between the plug and the head to reduce the protrusion of the plug for just this reason, on older engines with severe oil burning problems; this will cause the ignition of the fuel-air mixture to be less effective, but in such cases, this is of lesser significance.

Heat Range

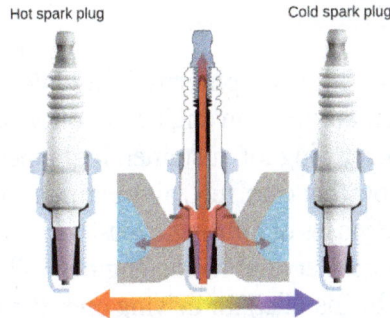

Hot spark plug   Cold spark plug

Construction of hot and cold spark plugs – a longer insulator tip makes the plug hotter

The operating temperature of a spark plug is the actual physical temperature at the tip of the spark plug within the running engine, normally between 500 °C and 800 °C. This is important because it determines the efficiency of plug self-cleaning and is determined by a number of factors, but primarily the actual temperature within the combustion chamber. There is no direct relationship between the actual operating temperature of the spark plug and spark voltage. However, the level of torque currently being produced by the engine will strongly influence spark plug operating temperature because the maximal temperature and pressure occur when the engine is operating near peak torque output (torque and rotational speed directly determine the power output). The temperature of the insulator responds to the thermal conditions it is exposed to in the combustion chamber, but not vice versa. If the tip of the spark plug is too hot, it can cause pre-ignition or sometimes detonation/knocking, and damage may occur. If it is too cold, electrically conductive deposits may form on the insulator, causing a loss of spark energy or the actual shorting-out of the spark current.

A spark plug is said to be "hot" if it is a better heat insulator, keeping more heat in the tip of the spark plug. A spark plug is said to be "cold" if it can conduct more heat out of the spark plug tip and lower the tip's temperature. Whether a spark plug is "hot" or "cold" is known as the heat range of the spark plug. The heat range of a spark plug is typically specified as a number, with some manufacturers using ascending numbers for hotter plugs, and others doing the opposite – using ascending numbers for colder plugs.

The heat range of a spark plug is affected by the construction of the spark plug: the types of materials used, the length of insulator and the surface area of the plug exposed within the combustion chamber. For normal use, the selection of a spark plug heat range is a balance between keeping the tip hot enough at idle to prevent fouling and cold enough at maximal power to prevent pre-ignition or engine knocking. By examining "hotter" and "cooler" spark plugs of the same manufacturer side by side, the principle involved can be very clearly seen; the cooler plugs have a more substantial ceramic insulator filling the gap between the center electrode and the shell, effectively allowing more heat to be carried off by the shell, while the hotter plugs have less ceramic material, so that the tip is more isolated from the body of the plug and retains heat better.

Heat from the combustion chamber escapes through the exhaust gases, the side walls of the cylinder and the spark plug itself. The heat range of a spark plug has only a minute effect on combustion chamber and overall engine temperature. A cold plug will not materially cool down an engine's running temperature. (A too hot plug may, however, indirectly lead to a runaway pre-ignition condition that *can* increase engine temperature.) Rather, the main effect of a "hot" or "cold" plug is to affect the temperature of the tip of the spark plug.

It was common before the modern era of computerized fuel injection to specify at least a couple of different heat ranges for plugs for an automobile engine; a hotter plug for cars that were mostly driven slowly around the city, and a colder plug for sustained high-speed highway use. This practice has, however, largely become obsolete now that cars' fuel/air mixtures and cylinder temperatures are maintained within a narrow range, for purposes of limiting emissions. Racing engines, however, still benefit from picking a proper plug heat range. Very old racing engines will sometimes have two sets of plugs, one just for starting and another to be installed for driving once the engine is warmed up.

Spark plug manufacturers use different numbers to denote heat range of their spark plugs.

Reading Spark Plugs

The spark plug's firing end will be affected by the internal environment of the combustion chamber. As the spark plug can be removed for inspection, the effects of combustion on the plug can be examined. An examination, or "reading" of the characteristic markings on the firing end of the spark plug can indicate conditions within the running engine. The spark plug tip will bear the marks as evidence of what is happening inside the engine. Usually there is no other way to know what is going on inside an engine running at peak power. Engine and spark plug manufacturers will publish information about the characteristic markings in spark plug reading charts. Such charts are useful for general use but are of almost no use in reading racing engine spark plugs, which is an entirely different matter.

A light brownish discoloration of the tip of the block indicates proper operation; other conditions may indicate malfunction. For example, a sandblasted look to the tip of the spark plug means persistent, light detonation is occurring, often unheard. The damage that is occurring to the tip of the spark plug is also occurring on the inside of the cylinder. Heavy detonation can cause outright breakage of the spark plug insulator and internal engine parts before appearing as sandblasted erosion but is easily heard. As another example, if the plug is too cold, there will be deposits on the nose of the plug. Conversely if the plug is too hot, the porcelain will be porous looking, almost like sugar. The material which seals the central electrode to the insulator will boil out. Sometimes the end of the plug will appear glazed, as the deposits have melted.

An idling engine will have a different impact on the spark plugs than one running at full throttle. Spark plug readings are only valid for the most recent engine operating conditions and running the engine under different conditions may erase or obscure characteristic marks previously left on the spark plugs. Thus, the most valuable information is gathered by running the engine at high speed and full load, immediately cutting the ignition off and stopping without idling or low speed operation and removing the plugs for reading.

Spark plug reading viewers, which are simply combined flashlight/magnifiers, are available to improve the reading of the spark plugs.

Two spark plug viewers

Indexing Spark Plugs

A matter of some debate is the "indexing" of plugs upon installation, usually only for high-performance or racing applications; this involves installing them so that the open area of the spark gap, not shrouded by the ground electrode, faces the center of the combustion chamber, towards the intake valve, rather than the wall. Some engine tuners believe that this will maximize the exposure of the fuel-air mixture to the spark, also ensuring that every combustion chamber is even in layout and therefore resulting in better ignition; however, others believe this is useful only to keep the ground electrode out of the way of the piston in ultra-high-compression engines if clearance is insufficient. In any event, this is accomplished by marking the location of the gap on the outside of the plug, installing it, and noting the direction in which the mark faces; then the plug is removed, and additional washers are added to change the orientation of the tightened plug. This must be done individually for each plug, as the orientation of the gap with respect to the threads of the shell is random. Some plugs are made with a non-random orientation of the gap and are usually marked as such by a suffix to the model number; typically these are specified by manufacturers of very small engines where the spark plug tip and electrodes form a significantly large part of the shape of the combustion chamber. The Honda Insight has indexed spark plugs from factory, with four different part numbers available corresponding to the different degrees of indexing to achieve most efficient combustion and maximal fuel efficiency.

Camshaft

Computer animation of a camshaft operating valves

A camshaft is a shaft to which a cam is fastened or of which a cam forms an integral part.

History

An early cam was built into Hellenistic water-driven automata from the 3rd century BC. The camshaft was later described in Turkey (Diyarbakır) by Al-Jazari in 1206. He employed it as part of his automata, water-raising machines, and water clocks such as the castle clock. The cam and camshaft later appeared in European mechanisms from at least the 14th century, or possibly earlier.

Uses

In internal combustion engines with pistons, the camshaft is used to operate poppet valves. It then consists of a cylindrical rod running the length of the cylinder bank with a number of oblong *lobes* protruding from it, one for each valve. The cam lobes force the valves open by pressing on the valve, or on some intermediate mechanism as they rotate.

Automotive

Materials

Camshafts can be made out of several types of material. These include:

Chilled iron castings: Commonly used in high volume production, chilled iron camshafts have a good wear resistance since the chilling process hardens them. Other elements are added to the iron before casting to make the material more suitable for its application.

Billet Steel: When a high quality camshaft or low volume production is required, engine builders and camshaft manufacturers choose steel billet. This is a much more time consuming process, and is generally more expensive than other methods. However the finished product is far superior. CNC lathes, CNC milling machines and CNC camshaft grinders will be used during production. Different types of steel bar can be used, one example being EN40b. When manufacturing a camshaft from EN40b, the camshaft will also be heat treated via gas nitriding, which changes the micro-structure of the material. It gives a surface hardness of 55-60 HRC. These types of camshafts can be used in high-performance engines.

Timing

The relationship between the rotation of the camshaft and the rotation of the crankshaft is of critical importance. Since the valves control the flow of the air/fuel mixture intake and exhaust gases, they must be opened and closed at the appropriate time during the stroke of the piston. For this reason, the camshaft is connected to the crankshaft either directly, via a gear mechanism, or indirectly via a belt or chain called a *timing belt* or *timing chain*. Direct drive using gears is unusual because of the cost. The frequently reversing torque caused by the slope of the cams tends to cause gear rattle which for an all-metal gear train requires further expense of a cam damper. Rolls-Royce V8 (1954) used gear drive as unlike chain it could be made silent and to last the life of the engine. Where gears are used in cheaper cars, they tend to be made from resilient fibre rather than metal, except in racing engines that have a high maintenance routine. Fibre gears have a short life span and must be replaced regularly, much like a timing belt. In some designs the camshaft also drives the distributor and the oil and fuel pumps. Some vehicles may have the power steering pump driven by the camshaft. With some early fuel injection systems, cams on the camshaft would operate the fuel injectors. Honda redesigned the VF750 from chain drive to gear drive VFR750 due to insurmountable problems with the VF750 Hi-Vo inverted chain drive.

A steel billet racing camshaft with noticeably broad lobes (very long duration)

An alternative used in the early days of OHC engines was to drive the camshaft(s) via a vertical shaft with bevel gears at each end. This system was, for example, used on the pre-WW1 Peugeot and Mercedes Grand Prix cars. Another option was to use a triple eccentric with connecting rods; these were used on certain W.O. Bentley-designed engines and also on the Leyland Eight.

In a two-stroke engine that uses a camshaft, each valve is opened once for every rotation of the crankshaft; in these engines, the camshaft rotates at the same speed as the crankshaft. In a four-stroke engine, the valves are opened only half as often; thus, two full rotations of the crankshaft occur for each rotation of the camshaft.

The timing of the camshaft can be advanced to produce better low RPM torque, or retarded for better high RPM power. Either of these moves the overall power produced by the engine down or up the RPM scale respectively. The amount of change is very little (usually < 5 deg), and affects valve to piston clearances. Refer this video https://www.youtube.com/watch?v=Hz1REougcfU

Duration

Duration is the number of crankshaft degrees of engine rotation during which the valve is off the seat. As a generality, greater duration results in more horsepower. The RPM at which peak horsepower occurs is typically increased as duration increases at the expense of lower rpm efficiency (torque).

Duration can often be confusing because manufacturers may select any lift point to advertise a camshaft's duration and sometimes will manipulate these numbers. The power and idle characteristics of a camshaft rated at .006" will be much different than one rated the same at .002".

Many performance engine builders gauge a race profile's aggressiveness by looking at the duration at .020", .050" and .200". The .020" number determines how responsive the motor will be and how much low end torque the motor will make. The .050" number is used to estimate where peak power will occur, and the .200" number gives an estimate of the power potential.

A secondary effect of increased duration is increasing *overlap*, which is the number of crankshaft degrees during which both intake and exhaust valves are off their seats. It is overlap which most affects idle quality, inasmuch as the "blow-through" of the intake charge which occurs during overlap reduces engine efficiency, and is greatest during low RPM operation. In reality, increasing a camshaft's duration typically increases the overlap event, unless one spreads lobe centers between intake and exhaust valve lobe profiles.

Lift

The camshaft "lift" is the resultant net rise of the valve from its seat. The further the valve rises from its seat the more airflow can be released, which is generally more beneficial. Greater lift has some limitations. Firstly, the lift is limited by the increased proximity of the valve head to the piston crown and secondly greater effort is required to move the valve's springs to higher state of compression. Increased lift can also be limited by lobe clearance in the cylinder head construction, so higher lobes may not necessarily clear the framework of the cylinder head casing. Higher valve lift can have the same effect as increased duration where valve overlap is less desirable.

Higher lift allows accurate timing of airflow; although even by allowing a larger volume of air to pass in the relatively larger opening, the brevity of the typical duration with a higher lift cam results in less airflow than with a cam with lower lift but more duration, all else being equal. On forced induction motors this higher lift could yield better results than longer duration, particularly on the intake side. Notably though, higher lift has more potential problems than increased duration, in particular as valve train rpm rises which can result in more inefficient running or loss of torque.

Cams that have too high a resultant valve lift, and at high rpm, can result in what is called "valve bounce", where the valve spring tension is insufficient to keep the valve following the cam at its apex. This could also be as a result of a very steep rise of the lobe and short duration, where the valve is effectively shot off the end of the cam rather than have the valve follow the cams' profile. This is typically what happens on a motor over rev. This is an occasion where the engine rpm exceeds the engine maximum design speed. The valve train is typically the limiting factor in determining the maximum rpm the engine can maintain either for a prolonged period or temporarily. Sometimes an over rev can cause engine failure where the valve stems become bent as a result of colliding with the piston crowns.

Position

Depending on the location of the camshaft, the cams operate the valves either directly or through a linkage of pushrods and rockers. Direct operation involves a simpler mechanism and leads to fewer failures, but requires the camshaft to be positioned at the top of the cylinders. In the past when engines were not as reliable as today this was seen as too much bother, but in modern gasoline engines the overhead cam system, where the camshaft is on top of the cylinder head, is quite common.

Number of Camshafts

While today some cheaper engines rely on a single camshaft per cylinder bank, which is known as a *single overhead camshaft* (SOHC), most modern engine designs (the *overhead-valve or OHV* engine being largely obsolete on passenger vehicles), are driven by a two camshafts per cylinder bank arrangement (one camshaft for the intake valves and another for the exhaust valves); such camshaft arrangement is known as a *double* or *dual overhead cam* (DOHC), thus, a V engine, which has two separate cylinder banks, may have four camshafts (colloquially known as a *quad-cam engine*).

More unusual is the *modern* W engine (also known as a 'VV' engine to distinguish itself from the pre-war W engines) that has *four* cylinder banks arranged in a "W" pattern with two pairs narrowly arranged with a 15-degree separation. Even when there are four cylinder banks (that would normally require a total of eight individual camshafts), the narrow-angle design allows the use of just four camshafts in total. For the Bugatti Veyron, which has a 16-cylinder W engine configuration, all the four camshafts are driving a total of 64 valves.

The overhead camshaft design adds more valvetrain components that ultimately incur in more complexity and higher manufacturing costs, but this is easily offset by many advantages over the older OHV design: multi-valve design, higher RPM limit and design freedom to better place valves, ignition (Spark-ignition engine) and intake/exhaust ports.

Maintenance

The rockers or cam followers sometimes incorporate a mechanism to adjust and set the valve play through manual adjustment, but most modern auto engines have hydraulic lifters, eliminating the need to adjust the valve lash at regular intervals as the valvetrain wears, and in particular the valves and valve seats in the combustion chamber.

Sliding friction between the surface of the cam and the cam follower which rides upon it is considerable. In order to reduce wear at this point, the cam and follower are both surface hardened, and modern lubricant motor oils contain additives specifically to reduce sliding friction. The lobes of the camshaft are usually slightly tapered, causing the cam followers or valve lifters to rotate slightly with each depression, and helping to distribute wear on the parts. The surfaces of the cam and follower are designed to "wear in" together, and therefore when either is replaced, the other should be as well to prevent excessive rapid wear. In some engines, the flat contact surfaces are replaced with rollers, which eliminate the sliding friction and wear but adds mass to the valvetrain.

Camshaft bearings are similar to crankshaft main bearings, being pressure-fed with oil. However, OHC camshaft bearings do not always have replaceable bearing shells, meaning that a new cylinder head is required if the bearings suffer wear due to insufficient or dirty oil.

Alternatives

In addition to mechanical friction, considerable force is required to overcome the valve springs used to close the engine's valves. This can amount to an estimated 25% of an engine's total output at idle, reducing overall efficiency. Some approaches to reclaiming this "wasted" energy include:

- Springless valves, like the desmodromic system employed today by Ducati

- Camless valvetrains using solenoids or magnetic systems have long been investigated by BMW and Fiat, and are currently being prototyped by Valeo and Ricardo

- The Wankel engine, a rotary engine which uses neither pistons nor valves, best known for being used by Mazda in the RX-7 and RX-8 sports cars.

- Koenigsegg has developed an electric valve actuator as a more fuel efficient and space saving alternative to the traditional camshaft.

Ignition Systems

In mechanically timed ignition systems, a separate camshaft is geared to the engine and operates a breaker that triggers a spark at the correct points in the combustion cycle.

Electrical

Before the advent of solid state electronics, *camshaft controllers* were used to control the speed of electric motors. A camshaft, driven by an electric motor or a pneumatic motor, was used to operate switches in sequence. By this means, resistors or tap changers were switched in or out of the circuit to vary the speed of the main motor. This system was widely used in electric multiple units.

Cam

A cam is a rotating or sliding piece in a mechanical linkage used especially in transforming rotary motion into linear motion or vice versa. It is often a part of a rotating wheel (e.g. an eccentric wheel) or shaft (e.g. a cylinder with an irregular shape) that strikes a lever at one or more points on its circular path. The cam can be a simple tooth, as is used to deliver pulses of power to a steam hammer, for example, or an eccentric disc or other shape that produces a smooth reciprocating (back and forth) motion in the *follower*, which is a lever making contact with the cam.

Overview

The cam can be seen as a device that rotates from circular to reciprocating (or sometimes oscillating) motion. A common example is the camshaft of an automobile, which takes the rotary motion of the engine and translates it into the reciprocating motion necessary to operate the intake and exhaust valves of the cylinders.

Plate Cam

The most commonly used cam is the plate cam (also *disc cam* or *radial cam*) which is cut out of a piece of flat metal or plate. Here, the follower moves in a plane perpendicular to the axis of rotation of the camshaft. Several key terms are relevant in such a construction of plate cams: base circle, prime circle (with radius equal to the sum of the follower radius and the base circle radius), pitch curve which is the radial curve traced out by applying the radial displacements away from the prime circle across all angles, and the lobe separation angle (LSA - the angle between two adjacent intake and exhaust cam lobes).

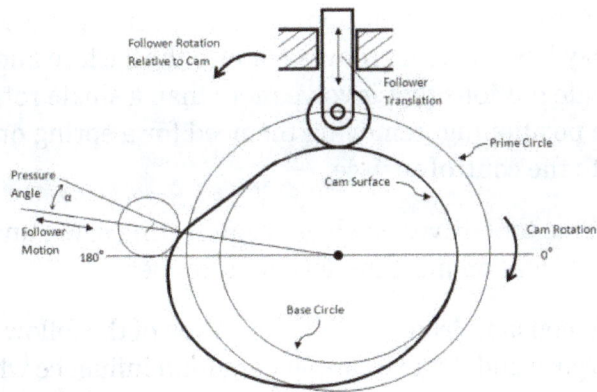

Fig. 3 Cam profile

The base circle is the smallest circle that can be drawn to the cam profile.

A once common, but now outdated, application of this type of cam was automatic machine tool programming cams. Each tool movement or operation was controlled directly by one or more cams. Instructions for producing programming cams and cam generation data for the most common makes of machine were included in engineering references well into the modern CNC era.

This type of cam is used in many simple electromechanical appliance controllers, such as dishwashers and clothes washing machines, to actuate mechanical switches that control the various parts.

Cylindrical Cam

Motorcycle transmission showing cylindrical cam with three followers.
Each follower controls the position of a shift fork.

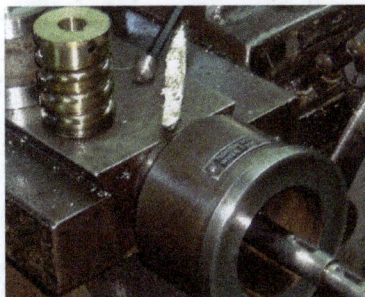

Constant lead barrel cam in an American Pacemaker lathe.
This cam is used to provide a repeatable cross slide setting when threading with a single-point tool.

A cylindrical cam or barrel cam is a cam in which the follower rides on the surface of a cylinder. In the most common type, the follower rides in a groove cut into the surface of a cylinder. These cams are principally used to convert rotational motion to linear motion parallel to the rotational axis of the cylinder. A cylinder may have several grooves cut into the surface and drive several followers. Cylindrical cams can provide motions that involve more than a single rotation of the cylinder and generally provide positive positioning, removing the need for a spring or other provision to keep the follower in contact with the control surface.

Applications include machine tool drives, such as reciprocating saws, and shift control barrels in sequential transmissions, such as on most modern motorcycles.

A special case of this cam is constant lead, where the position of the follower is linear with rotation, as in a lead screw. The purpose and detail of implementation influence whether this application is called a cam or a screw thread, but in some cases, the nomenclature may be ambiguous.

Cylindrical cams may also be used to reference an output to two inputs, where one input is rotation of the cylinder, and the second is position of the follower axially along the cam. The output is radial to the cylinder. These were once common for special functions in control systems, such as fire control mechanisms for guns on naval vessels and mechanical analog computers.

An example of a cylindrical cam with two inputs is provided by a duplicating lathe, an example of which is the Klotz axe handle lathe, which cuts an axe handle to a form controlled by a pattern acting as a cam for the lathe mechanism.

Face Cam

A face cam produces motion by using a follower riding on the face of a disk. The most common type has the follower ride in a slot so that the captive follower produces radial motion with positive positioning without the need for a spring or other mechanism to keep the follower in contact with the control surface. A face cam of this type generally has only one slot for a follower on each face. In some applications, a single element, such as a gear, a barrel cam, or other rotating element with a flat face, may do duty as a face cam in addition to other purposes.

Face cams may provide repetitive motion with a groove that forms a closed curve, or may provide function generation with a stopped groove. Cams used for function generation may have grooves that require several revolutions to cover the complete function, and in this case, the function generally needs to be invertible so that the groove does not self intersect, and the function output value must differ enough at corresponding rotations that there is sufficient material separating the adjacent groove segments. A common form is the constant lead cam, where displacement of the follower is linear with rotation, such as the scroll plate in a scroll chuck. Non-invertible functions, which require the groove to self-intersect, can be implemented using special follower designs.

Sash window lock, traditional cam style, for double-hung sash window

A variant of the face cam provides motion parallel to the axis of cam rotation. A common example is the traditional sash window lock, where the cam is mounted to the top of the lower sash, and the follower is the hook on the upper sash. In this application, the cam is used to provide mechanical advantage in forcing the window shut, and also provides a self-locking action, like some worm gears, due to friction.

Face cams may also be used to reference a single output to two inputs, typically where one input is rotation of the cam and the other is radial position of the follower. The output is parallel to the axis of the cam. These were once common is mechanical analog computation and special functions in control systems.

A face cam that implements three outputs for a single rotational input is the stereo phonograph, where a relatively constant lead groove guides the stylus and tone arm unit, acting as either a rocker-type (tone arm) or linear (linear tracking turntable) follower, and the stylus alone acting as the follower for two orthogonal outputs to representing the audio signals. These motions are in a plane radial to the rotation of the record and at angles of 45 degrees to the plane of the disk (normal to the groove faces). The position of the tone arm was used by some turntables as a control input, such as to turn the unit off or to load the next disk in a stack, but was ignored in simple units.

Heart Shaped Cam

This type of cam, in the form of a symmetric heart symbol, is used to return a shaft holding the cam to a set position by pressure from a roller. They were used for example on early models of Post Office Master clocks to synchronise the clock time with Greenwich Mean Time when the activating follower was pressed onto the cam automatically via a signal from an accurate time source.

Snail Drop Cam

This type of cam was used for example in mechanical time keeping clocking-in clocks to drive the day advance mechanism at precisely midnight and consisted of a follower being raised over 24 hours by the cam in a spiral path which terminated at a sharp cut off at which the follower would drop down and activate the day advance. Where timing accuracy is required as in clocking-in clocks these were typically ingeniously arranged to have a roller cam follower to raise the drop weight for most of its journey to near its full height, and only for the last portion of its travel for the weight to be taken over and supported by a solid follower with a sharp edge. This ensured that the weight dropped at a precise moment, enabling accurate timing. This was achieved by the use of two snail cams mounted coaxially with the roller initially resting on one cam and the final solid follower on the other but not in contact with its cam profile. Thus the roller cam was initially carried the weight, until at the final portion of the run the profile of the non-roller cam rose more than the other causing the solid follower to take the weight.

Linear Cam

A linear cam is one in which the cam element moves in a straight line rather than rotates. The cam element is often a plate or block, but may be any cross section. The key feature is that the input is a linear motion rather than rotational. The cam profile may be cut into one or more edges of a plate or block, may be one or more slots or grooves in the face of an element, or may even be a surface profile for a cam with more than one input. The development of a linear cam is similar to, but not identical to, that of a rotating cam.

A common example of a linear cam is a key for a pin tumbler lock. The pins act as the followers. This behavior is exemplified when the key is duplicated in a key duplication machine, where the original key acts as a control cam for cutting the new key.

Key duplicating machine. The original key (mounted in the left hand holder) acts as a linear cam to control the cut depth for the duplicate.

History

An early cam was built into Hellenistic water-driven automata from the 3rd century BC. Cams were later employed by Al-Jazari, who used them in his own automata. The cam and camshaft appeared in European mechanisms from the 14th century.

Gudgeon Pin

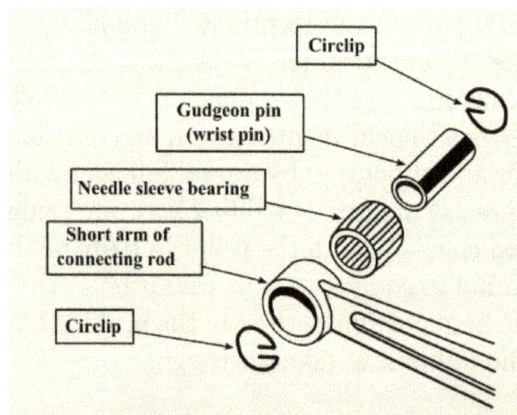

Gudgeon pin connection at connecting rod. Gudgeon pin fits into gudgeons inside piston.

In internal combustion engines, the gudgeon pin (UK, wrist pin US) connects the piston to the connecting rod and provides a bearing for the connecting rod to pivot upon as the piston moves. In very early engine designs (including those driven by steam and also many very large stationary or marine engines), the gudgeon pin is located in a sliding crosshead that connects to the piston via a rod. A gudgeon is a pivot or journal. The origin of the word gudgeon is the Middle English word gojoun, which originated from the Middle French word goujon. Its first known use was in the 15th century.

Overview

The gudgeon pin is typically a forged short hollow rod made of a steel alloy of high strength and hardness that may be physically separated from both the connecting rod and piston or crosshead. The design of the gudgeon pin, especially in the case of small, high-revving automotive engines is challenging. The gudgeon pin has to operate under some of the highest temperatures experienced in the engine, with difficulties in lubrication due to its location, while remaining small and light so as to fit into the piston diameter and not unduly add to the reciprocating mass. The requirements for lightness and compactness demand a small diameter rod that is subject to heavy shear and bending loads, with some of the highest pressure loadings of any bearing in the whole engine. To overcome these problems, the materials used to make the gudgeon pin and the way it is manufactured are amongst the most highly engineered of any mechanical component found in internal combustion engines.

Design Options

Gudgeon pins use two broad design configurations: semi-floating and fully floating.

Semi-floating

In the semi-floating configuration, the pin is usually fixed relative to the piston by an interference fit with the journal in the piston. (This replaced the earlier set screw method.) The connecting rod small end bearing thus acts as the bearing alone. In this configuration, only the small end bearing requires a bearing surface, if any. If needed, this is provided by either electroplating the small end bearing journal with a suitable metal, or more usually by inserting a sleeve bearing or needle bearing into the eye of the small end, which has an interference fit with the aperture of the small end. During overhaul, it is usually possible to replace this bearing sleeve if it is badly worn. The reverse configuration, fixing the gudgeon pin to the connecting rod instead of to the piston, is implemented using an interference fit with the small end eye instead, with the gudgeon pin journals in the piston functioning as bearings. This arrangement is usually more difficult to manufacture and service because two bearing surfaces or inserted sleeves complicate the design. In addition, the pin must be precisely set so that the small end eye is central. Because of thermal expansion considerations, this arrangement was more usual for single-cylinder engines as opposed to multiple cylinder engines with long cylinder blocks and crankcases, until precision manufacturing became more commonplace.

Fully floating

In the fully floating configuration, a bearing surface is created both between the small end eye and gudgeon pin and the journal in the piston. The gudgeon pins are usually secured with circlips. No interference fit is used in any instance and the pin 'floats' entirely on bearing surfaces. The average rubbing speed of each of the three bearings is halved and the load is shared across a bearing that is usually about three times the length of the semi-floating design with an interference fit with the piston.

Connecting Rod

In a reciprocating piston engine, the connecting rod or conrod connects the piston to the crank or crankshaft. Together with the crank, they form a simple mechanism that converts reciprocating motion into rotating motion.

Connecting rods may also convert rotating motion into reciprocating motion. Historically, before the development of engines, they were first used in this way.

As a connecting rod is rigid, it may transmit either a push or a pull and so the rod may rotate the crank through both halves of a revolution, i.e. piston pushing and piston pulling. Earlier mechanisms, such as chains, could only pull. In a few two-stroke engines, the connecting rod is only required to push.

Today, connecting rods are best known through their use in internal combustion piston engines, such as automotive engines. These are of a distinctly different design from earlier forms of connecting rods, used in steam engines and steam locomotives.

History

Scheme of the Roman Hierapolis sawmill, the earliest known machine to combine a connecting rod with a crank.

The earliest evidence for a connecting rod appears in the late 3rd century AD Roman Hierapolis sawmill. It also appears in two 6th century Eastern Roman saw mills excavated at Ephesus and Gerasa. The crank and connecting rod mechanism of these Roman watermills converted the rotary motion of the waterwheel into the linear movement of the saw blades.

Sometime between 1174 and 1206, the Arab inventor and engineer Al-Jazari described a machine which incorporated the connecting rod with a crankshaft to pump water as part of a water-raising machine, but the device was unnecessarily complex indicating that he still did not fully understand the concept of power conversion.

In Renaissance Italy, the earliest evidence of a – albeit mechanically misunderstood – compound crank and connecting-rod is found in the sketch books of Taccola. A sound understanding of the motion involved is displayed by the painter Pisanello (d. 1455) who showed a piston-pump driven by a water-wheel and operated by two simple cranks and two connecting-rods.

By the 16th century, evidence of cranks and connecting rods in the technological treatises and artwork of Renaissance Europe becomes abundant; Agostino Ramelli's *The Diverse and Artifactitious Machines* of 1588 alone depicts eighteen examples, a number which rises in the *Theatrum Machinarum Novum* by Georg Andreas Böckler to 45 different machines.

Steam Engines

Beam engine, with twin connecting rods (almost vertical) between the horizontal beam and the flywheel cranks

The first steam engines, Newcomen's atmospheric engine, was single-acting: its piston only did work in one direction and so these used a chain rather than a connecting rod. Their output rocked back and forth, rather than rotating continuously.

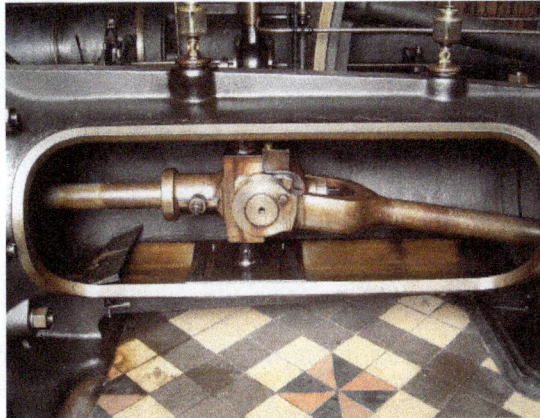

Crosshead of a stationary steam engine: piston rod to the left, connecting rod to the right

Steam engines after this are usually double-acting: their internal pressure works on each side of the piston in turn. This requires a seal around the piston rod and so the hinge between the piston and connecting rod is placed outside the cylinder, in a large sliding bearing block called a crosshead.

Steam locomotive rods, the large angled rod being the connecting rod

In a steam locomotive, the crank pins are usually mounted directly on one or more pairs of driving wheels, and the axle of these wheels serves as the crankshaft. The connecting rods (also called the main rods *in US practice*), run between the crank pins and crossheads, where they connect to the piston rods. Crossheads or trunk guides are also used on large diesel engines manufactured for marine service. (The similar rods between driving wheels are called coupling rods *in British practice*.)

The connecting rods of smaller steam locomotives are usually of rectangular cross-section but, on small locomotives, marine-type rods of circular cross-section have occasionally been used. Stephen Lewin, who built both locomotive and marine engines, was a frequent user of round rods. Gresley's A4 Pacifics, such as *Mallard*, had an alloy steel connecting rod in the form of an I-beam with a web that was only 0.375 in (9.53 mm) thick.

On Western Rivers steamboats, the connecting rods are properly called pitmans, and are sometimes incorrectly referred to as pitman arms.

Internal Combustion Engines

Three different connecting rods, of which the left and the aluminum center, the connecting rod to the right (for endothermic engine) in steel, the left connecting rod (for endothermic engine) has the modular head and the foot equipped with a bushing, the central rod has the oil drip rod equipped with pats

In modern automotive internal combustion engines, the connecting rods are most usually made of steel for production engines, but can be made of T6-2024 and T651-7075 aluminum alloys (for lightness and the ability to absorb high impact at the expense of durability) or titanium (for a combination of lightness with strength, at higher cost) for high-performance engines, or of cast iron for applications such as motor scooters. They are not rigidly fixed at either end, so that the angle between the connecting rod and the piston can change as the rod moves up and down and rotates around the crankshaft. Connecting rods, especially in racing engines, may be called "billet" rods, if they are machined out of a solid billet of metal, rather than being cast or forged.

Small End and Big End

The small end attaches to the piston pin, gudgeon pin or wrist pin, which is currently most often press fit into the connecting rod but can swivel in the piston, a "floating wrist pin" design. The big end connects to the bearing journal on the crank throw, in most engines running on replaceable bearing shells accessible via the *connecting rod bolts* which hold the bearing "cap" onto the big end. Typically there is a pinhole bored through the bearing and the big end of the connecting rod so that pressurized lubricating motor oil squirts out onto the thrust side of the cylinder wall to lubricate the travel of the pistons and piston rings. Most small two-stroke engines and some single cylinder four-stroke engines avoid the need for a pumped lubrication system by using a rolling-element bearing instead, however this requires the crankshaft to be pressed apart and then back together in order to replace a connecting rod.

Engine Wear and Rod Length

A major source of engine wear is the sideways force exerted on the piston through the connecting rod by the crankshaft, which typically wears the cylinder into an oval cross-section rather than circular, making it impossible for piston rings to correctly seal against the cylinder walls. Geometrically, it can be seen that longer connecting rods will reduce the amount of this sideways force, and therefore lead to longer engine life. However, for a given engine block, the sum of the length of the connecting rod plus the piston stroke is a fixed number, determined by the fixed distance between the crankshaft axis and the top of the cylinder block where the cylinder head fastens; thus, for a given cylinder block longer stroke, giving greater engine displacement and power, requires a shorter connecting rod (or a piston with smaller compression height), resulting in accelerated cylinder wear.

Stress and Failure

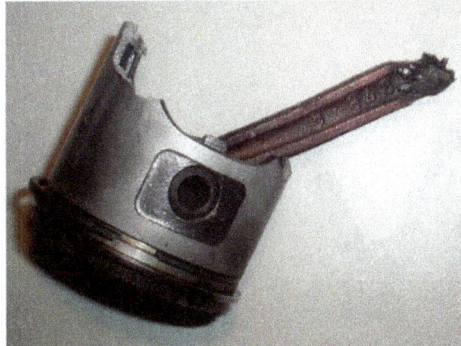

Failure of a connecting rod is one of the most common causes of catastrophic engine failure.

The connecting rod is under tremendous stress from the reciprocating load represented by the piston, actually stretching and being compressed with every rotation, and the load increases as the square of the engine speed increase. Failure of a connecting rod, usually called throwing a rod, is one of the most common causes of catastrophic engine failure in cars, frequently putting the broken rod through the side of the crankcase and thereby rendering the engine irreparable; it can result from fatigue near a physical defect in the rod, lubrication failure in a bearing due to faulty maintenance, or from failure of the rod bolts from a defect, improper tightening or over-revving of the engine. In an unmaintained, dirty environment, a water or chemical emulsifies with the oil that lubricates the bearing and causes the bearing to fail. Re-use of rod bolts is a common practice as long as the bolts meet manufacturer specifications. Despite their frequent occurrence on televised competitive automobile events, such failures are quite rare on production cars during normal daily driving. This is because production auto parts have a much larger factor of safety, and often more systematic quality control.

High-performance Engines

When building a high-performance engine, great attention is paid to the connecting rods, eliminating stress risers by such techniques as grinding the edges of the rod to a smooth radius, shot peening to induce compressive surface stresses (to prevent crack initiation), balancing all connecting rod/piston assemblies to the same weight and Magnafluxing to reveal otherwise invisible small cracks which would cause the rod to fail under stress. In addition, great care is taken to torque the connecting rod bolts to the exact value specified; often these bolts must be replaced rather than reused. The big end of the rod is fabricated as a unit and cut or cracked in two to establish precision fit around the big end bearing shell. Therefore, the big end "caps" are not interchangeable between connecting rods, and when rebuilding an engine, care must be taken to ensure that the caps of the different connecting rods are not mixed up. Both the connecting rod and its bearing cap are usually embossed with the corresponding position number in the engine block.

Powder Metallurgy

Recent engines such as the Ford 4.6 litre engine and the Chrysler 2.0 litre engine, have connecting rods made using powder metallurgy, which allows more precise control of size and weight with less machining and less excess mass to be machined off for balancing. The cap is then separated from the rod by a fracturing process, which results in an uneven mating surface due to the grain of the powdered metal. This ensures that upon reassembly, the cap will be perfectly positioned with respect to the rod, compared to the minor misalignments which can occur if the mating surfaces are both flat.

Compound Rods

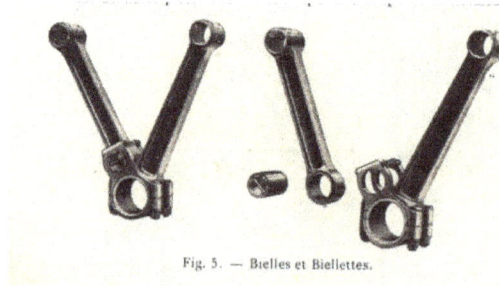

Fig. 5. — Bielles et Biellettes.

Articulated connecting rods

Many-cylinder multi-bank engines such as a V12 layout have little space available for many connecting rod journals on a limited length of crankshaft. This is a difficult compromise to solve and its consequence has often led to engines being regarded as failures (Sunbeam Arab, Rolls-Royce Vulture).

The simplest solution, almost universal in road car engines, is to use simple rods where cylinders from both banks share a journal. This requires the rod bearings to be *narrower*, increasing bearing load and the risk of failure in a high-performance engine. This also means the opposing cylinders are not exactly in line with each other.

In certain engine types, master/slave rods are used rather than the simple type shown in the picture above. The master rod carries one or more ring pins to which are bolted the much smaller big ends of slave rods on other cylinders. Certain designs of V engines use a master/slave rod for each pair of opposite cylinders. A drawback of this is that the stroke of the subsidiary rod is slightly shorter than the master, which increases vibration in a vee engine, catastrophically so for the Sunbeam Arab.

Radial engines typically have a master rod for one cylinder and multiple slave rods for all the other cylinders in the same bank.

BMW 132 radial aero engine rods

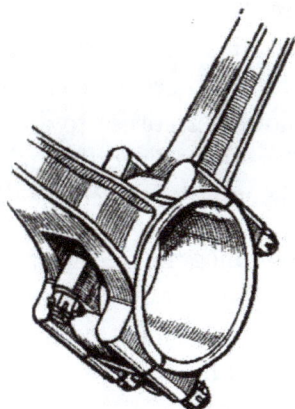

Fork and blade rods

The usual solution for high-performance aero-engines is a "forked" connecting rod. One rod is split in two at the big end and the other is thinned to fit into this fork. The journal is still shared between cylinders. The Rolls-Royce Merlin used this "fork-and-blade" style. A common arrangement for forked rods is for the fork rod to have a single wide bearing sleeve that spans the whole width of the rod, including the central gap. The blade rod then runs, not directly on the crankpin, but on the outside of this sleeve. The two rods do not rotate relative to each other, merely oscillate back and forth, so this bearing is relatively lightly loaded and runs as a much lower surface speed. However the bearing movement also becomes reciprocating rather than continuously rotating, which is a more difficult problem for lubrication.

A likely candidate for an extreme example of compound articulated rod design could be the complex German 24-cylinder Junkers Jumo 222 aviation engine, meant to have — unlike an X-engine layout with 24 cylinders, possessing six cylinders per bank — only four cylinders per bank, and *six* banks of cylinders, all liquid-cooled with five "slave" rods pinned to one master rod, for each "layer" of cylinders in its design. After building nearly 300 test examples in several different displacements, the Junkers firm's complex Jumo 222 engine turned out to be a production failure for the more advanced combat aircraft of the Third Reich's Luftwaffe which required aviation powerplants of over 1,500 kW (2,000 PS) output apiece.

Crankshaft

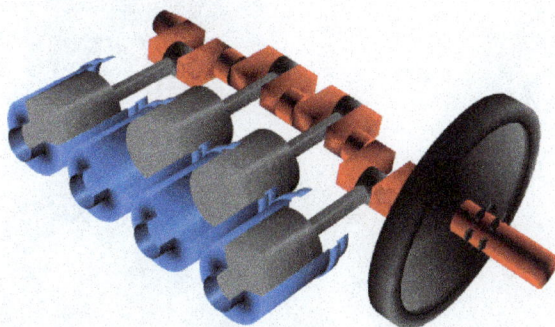

Flat-plane crankshaft (red), pistons (gray) in their cylinders (blue), and flywheel (black)

A crankshaft—related to *crank*—is a mechanical part able to perform a conversion between reciprocating motion and rotational motion. In a reciprocating engine, it translates reciprocating motion of the piston into rotational motion; whereas in a reciprocating compressor, it converts the rotational motion into reciprocating motion. In order to do the conversion between two motions, the crankshaft has "crank throws" or "crankpins", additional bearing surfaces whose axis is offset from that of the crank, to which the "big ends" of the connecting rods from each cylinder attach.

It is typically connected to a flywheel to reduce the pulsation characteristic of the four-stroke cycle, and sometimes a torsional or vibrational damper at the opposite end, to reduce the torsional vibrations often caused along the length of the crankshaft by the cylinders farthest from the output end acting on the torsional elasticity of the metal.

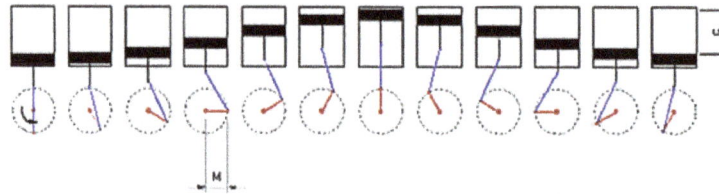

Schematic of operation of a crank mechanism

History

Roman Empire

A Roman iron crank of yet unknown purpose dating to the 2nd century AD was excavated in Augusta Raurica, Switzerland. The 82.5 cm long piece has fitted to one end a 15 cm long bronze handle, the other handle being lost.

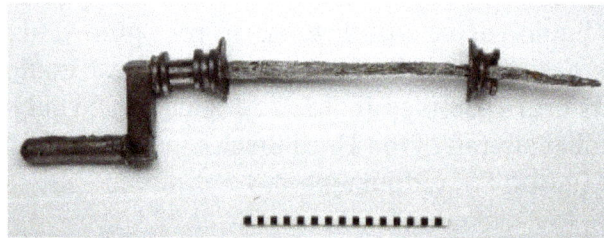

Roman crank dated to the 2nd century AD. The right handle is lost.

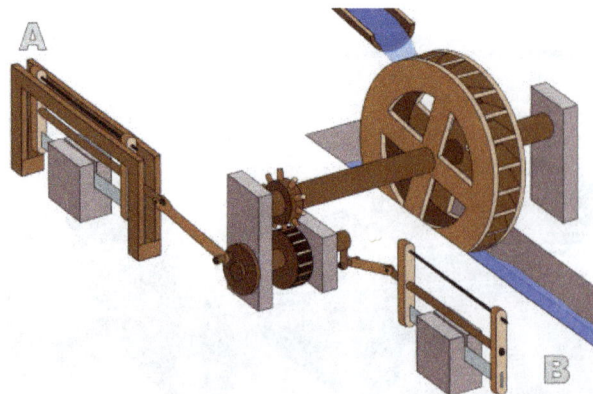

Roman Hierapolis sawmill from the 3rd century AD, the earliest known machine to combine a crank with a connecting rod.

The earliest evidence, anywhere in the world, for a crank and connecting rod in a machine appears in the late Roman Hierapolis sawmill from the 3rd century AD and two Roman stone sawmills at Gerasa, Roman Syria, and Ephesus, Asia Minor (both 6th century AD). On the pediment of the Hierapolis mill, a waterwheel fed by a mill race is shown transmitting power through a gear train to two frame saws, which cut rectangular blocks by way of some kind of connecting rods and, through mechanical necessity, cranks. The accompanying inscription is in Greek.

The crank and connecting rod mechanisms of the other two archaeologically attested sawmills worked without a gear train. In ancient literature, we find a reference to the workings of water-powered marble saws close to Trier, now Germany, by the late 4th century poet Ausonius; about the same time, these mill types seem also to be indicated by the Christian saint Gregory of Nyssa from Anatolia, demonstrating a diversified use of water-power in many parts of the Roman Empire. The three finds push back the date of the invention of the crank and connecting rod back by a full millennium; for the first time, all essential components of the much later steam engine were assembled by one technological culture:

With the crank and connecting rod system, all elements for constructing a steam engine (invented in 1712) — Hero's aeolipile (generating steam power), the cylinder and piston (in metal force pumps), non-return valves (in water pumps), gearing (in water mills and clocks) — were known in Roman times.

Medieval East

Al-Jazari (1136–1206) described a crank and connecting rod system in a rotating machine in two of his water-raising machines. His twin-cylinder pump incorporated a crankshaft, though the device was unnecessarily complex.

In China, the potential of the crank of converting circular motion into reciprocal one never seems to have been fully realized, and the crank was typically absent from such machines until the turn of the 20th century.

Medieval Europe

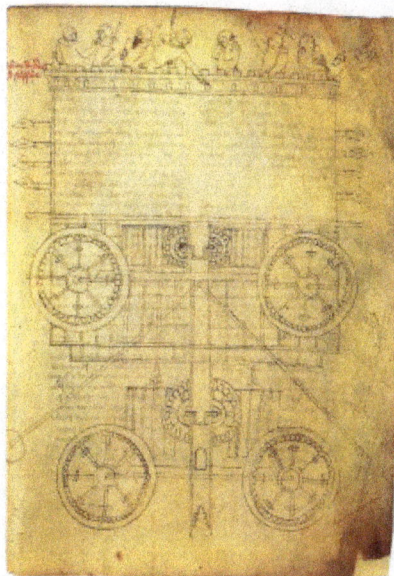

Vigevano's war carriage

The Italian physician Guido da Vigevano (c. 1280–1349), planning for a new crusade, made illustrations for a paddle boat and war carriages that were propelled by manually turned compound cranks and gear wheels (center of image). The Luttrell Psalter, dating to around 1340, describes a grindstone rotated by two cranks, one at each end of its axle; the geared hand-mill, operated either with one or two cranks, appeared later in the 15th century;

Renaissance Europe

15th century paddle-wheel boat whose paddles are turned by single-throw crankshafts
(Anonymous of the Hussite Wars)

The first depictions of the compound crank in the carpenter's brace appear between 1420 and 1430 in various northern European artwork. The rapid adoption of the compound crank can be traced in the works of the Anonymous of the Hussite Wars, an unknown German engineer writing on the state of the military technology of his day: first, the connecting-rod, applied to cranks, reappeared, second, double compound cranks also began to be equipped with connecting-rods and third, the flywheel was employed for these cranks to get them over the 'dead-spot'.

In Renaissance Italy, the earliest evidence of a compound crank and connecting-rod is found in the sketch books of Taccola, but the device is still mechanically misunderstood. A sound grasp of the crank motion involved is demonstrated a little later by Pisanello, who painted a piston-pump driven by a water-wheel and operated by two simple cranks and two connecting-rods.

Water-raising pump powered by crank and connecting rod mechanism (Georg Andreas Böckler, 1661)

One of the drawings of the Anonymous of the Hussite Wars shows a boat with a pair of paddle-wheels at each end turned by men operating compound cranks (see above). The concept was much improved by the Italian Roberto Valturio in 1463, who devised a boat with five sets, where the parallel cranks are all joined to a single power source by one connecting-rod, an idea also taken up by his compatriot Francesco di Giorgio.

Crankshafts were also described by Konrad Kyeser (d. 1405), Leonardo da Vinci (1452–1519) and a Dutch "farmer" by the name Cornelis Corneliszoon van Uitgeest in 1592. His wind-powered sawmill used a crankshaft to convert a windmill's circular motion into a back-and-forward motion powering the saw. Corneliszoon was granted a patent for his crankshaft in 1597.

From the 16th century onwards, evidence of cranks and connecting rods integrated into machine design becomes abundant in the technological treatises of the period: Agostino Ramelli's *The Diverse and Artifactitious Machines* of 1588 alone depicts eighteen examples, a number that rises in the *Theatrum Machinarum Novum* by Georg Andreas Böckler to 45 different machines, one third of the total.

Internal Combustion Engines

Crankshaft, pistons and connecting rods for a typical internal combustion engine

MAN marine crankshaft for 6cyl marine diesel applications. Note locomotive on left for size reference

Large engines are usually multicylinder to reduce pulsations from individual firing strokes, with more than one piston attached to a complex crankshaft. Many small engines, such as those found in mopeds or garden machinery, are single cylinder and use only a single piston, simplifying crankshaft design.

A crankshaft is subjected to enormous stresses, potentially equivalent of several tonnes of force. The crankshaft is connected to the fly-wheel (used to smooth out shock and convert energy to torque), the engine block, using bearings on the main journals, and to the pistons via their respective con-rods. An engine loses up to 75% of its generated energy in the form of friction, noise and vibration in the crankcase and piston area. The remaining losses occur in the valvetrain (timing chains, belts, pulleys, camshafts, lobes, valves, seals etc.) heat and blow by.

Bearings

The crankshaft has a linear axis about which it rotates, typically with several bearing journals riding on replaceable bearings (the main bearings) held in the engine block. As the crankshaft undergoes a great deal of sideways load from each cylinder in a multicylinder engine, it must be supported by several such bearings, not just one at each end. This was a factor in the rise of V8 engines, with their shorter crankshafts, in preference to straight-8 engines. The long crankshafts of the latter suffered from an unacceptable amount of flex when engine designers began using higher compression ratios and higher rotational speeds. High performance engines often have more main bearings than their lower performance cousins for this reason.

Piston Stroke

The distance the axis of the crank throws from the axis of the crankshaft determines the piston stroke measurement, and thus engine displacement. A common way to increase the low-speed torque of an engine is to increase the stroke, sometimes known as "shaft-stroking." This also increases the reciprocating vibration, however, limiting the high speed capability of the engine. In compensation, it improves the low speed operation of the engine, as the longer intake stroke through smaller valve(s) results in greater turbulence and mixing of the intake charge. Most modern high speed production engines are classified as "over square" or short-stroke, wherein the stroke is less than the diameter of the cylinder bore. As such, finding the proper balance between shaft-stroking speed and length leads to better results.

Engine Configuration

The configuration, meaning the number of pistons and their placement in relation to each other leads to straight, V or flat engines. The same basic engine block can sometimes be used with different crankshafts, however, to alter the firing order. For instance, the 90° V6 engine configuration, in older days sometimes derived by using six cylinders of a V8 engine with a 3 throw crankshaft, produces an engine with an inherent pulsation in the power flow due to the "gap" between the firing pulses alternates between short and long pauses because the 90 degree engine block does not correspond to the 120 degree spacing of the crankshaft. The same engine, however, can be made to provide evenly spaced power pulses by using a crankshaft with an individual crank throw for each cylinder, spaced so that the pistons are actually phased 120° apart, as in the GM 3800 engine. While most production V8 engines use four crank throws spaced 90° apart, high-performance

V8 engines often use a "flat" crankshaft with throws spaced 180° apart, essentially resulting in two straight four engines running on a common crankcase. The difference can be heard as the flat-plane crankshafts result in the engine having a smoother, higher-pitched sound than cross-plane (for example, IRL IndyCar Series compared to NASCAR Sprint Cup Series, or a Ferrari 355 compared to a Chevrolet Corvette). This type of crankshaft was also used on early types of V8 engines.

Engine Balance

For some engines it is necessary to provide counterweights for the reciprocating mass of each piston and connecting rod to improve engine balance. These are typically cast as part of the crankshaft but, occasionally, are bolt-on pieces. While counter weights add a considerable amount of weight to the crankshaft, it provides a smoother running engine and allows higher RPM levels to be reached.

Flying Arms

Crankshaft with flying arms (the boomerang-shaped link between the visible crankpins)

In some engine configurations, the crankshaft contains direct links between adjacent crankpins, without the usual intermediate main bearing. These links are called *flying arms*. This arrangement is sometimes used in V6 and V8 engines as it enables the engine to be designed with different V angles than what would otherwise be required to create an even firing interval, while still using fewer main bearings than would normally be required with a single piston per crankthrow. This arrangement reduces weight and engine length at the expense of less crankshaft rigidity.

Rotary Aircraft Engines

Some early aircraft engines were a rotary engine design, where the crankshaft was fixed to the airframe and instead the cylinders rotated with the propeller.

Radial Engines

The radial engine is a reciprocating type internal combustion engine configuration in which the cylinders point outward from a central crankshaft like the spokes of a wheel. It resembles a stylized star when viewed from the front, and is called a "star engine" (German Sternmotor, French Moteur en étoile) in some languages. The radial configuration was very commonly used in aircraft engines before turbine engines became predominant.

Construction

Continental engine marine crankshafts, 1942

Crankshafts can be monolithic (made in a single piece) or assembled from several pieces. Monolithic crankshafts are most common, but some smaller and larger engines use assembled crankshafts.

Forging and Casting

Forged crankshaft

Crankshafts can be forged from a steel bar usually through roll forging or cast in ductile steel. Today more and more manufacturers tend to favor the use of forged crankshafts due to their lighter weight, more compact dimensions and better inherent damping. With forged crankshafts, vanadium microalloyed steels are mostly used as these steels can be air cooled after reaching high strengths without additional heat treatment, with exception to the surface hardening of the bearing surfaces. The low alloy content also makes the material cheaper than high alloy steels. Carbon steels are also used, but these require additional heat treatment to reach the desired properties. Iron crankshafts are today mostly found in cheaper production engines (such as those found in the Ford Focus diesel engines) where the loads are lower. Some engines also use cast iron crankshafts for low output versions while the more expensive high output version use forged steel.

Machining

Crankshafts can also be machined out of a billet, often a bar of high quality vacuum remelted steel. Though the fiber flow (local inhomogeneities of the material's chemical composition generated during casting) doesn't follow the shape of the crankshaft (which is undesirable), this is usually not a problem since higher quality steels, which normally are difficult to forge, can be used. These crankshafts tend to be very expensive due to the large amount of material that must be removed with lathes and milling machines, the high material cost, and the additional heat treatment required. However, since no expensive tooling is needed, this production method allows small production runs without high costs.

In an effort to reduce costs, used crankshafts may also be machined. A good core may often be easily reconditioned by a crankshaft grinding process. Severely damaged crankshafts may also be repaired with a welding operation, prior to grinding, that utilizes a submerged arc welding machine. To accommodate the smaller journal diameters a ground crankshaft has, and possibly an over-sized thrust dimension, undersize engine bearings are used to allow for precise clearances during operation.

Machining or remanufacturing crankshafts are precision machined to exact tolerances with no odd size crankshaft bearings or journals. Thrust surfaces are micro-polished to provide precise surface finishes for smooth engine operation and reduced thrust bearing wear. Every journal is inspected and measured with critical accuracy. After machining, oil holes are chamfered to improve lubrication and every journal polished to a smooth finish for long bearing life. Remanufactured crankshafts are thoroughly cleaned with special emphasis to flushing and brushing out oil passages to remove any contaminants. Typically there are 23 steps to re-manufacturing a crankshaft which are as follows:

Step 1: Industrial Cleaning

The first step in the industrial crankshaft remanufacturing process is cleaning the entire crankshaft. Machine shops soak the rebuilt crankshafts in a hot tank and use a power washing station on the overall shaft as needed. Next machinists then wire brush all oil holes to remove caked on residue and other substances.

Step 2: Magnetic Particle Inspection

The second step in the crankshaft remanufacturing process is using a magnetic particle inspection method to check for cracks. The crankshaft is maganitized and sprayed with a iron oxide powder which, under blacklight conditions, makes any cracks or imperfections visible. All remanufactured crankshafts are checked for imperfections before proceeding forward in the manufacturing process.

Step 3: Check Counterweights

The machine shop then removes and cleans the counterweights. The production facility then checks the counterweights to make sure they are tight. If the counterweights are loose a technician then replaces all of the counterweight bolts. Counterweights are inspected for cracks before being replaced or retightened. In step sixteen the machinist re-installs the counterweights back into the rebuilt crankshafts.

Step 4: Check Crankshaft Bearings and Straightness

The machinist then inspects the entire incoming remanufactured crankshaft for damage and determines the size of the journals and mains. Next the machinist checks the hardness of the mains and journals. It is crucial to also inspect the crankshaft bearings and check the straightness of the overall crankshaft. Re-straightening the industrial crankshaft if not up to OEM standards occurs in step seven. Veteran machine shops typically do not re-straighten the rebuilt crankshafts until a quality control technician checks the bolt holes and seals the surface for divots.

Step 5: Check Bolt Holes

The technician checks the keyway, nose, bolt holes and seals the surface for non-conformities. Usually machine shops will tap bolt holes up to but not more than ½" on all remanufactured crankshafts.

Step 6: Stamp Counterweight Webbing

The rebuild team next stamps the counterweights & webbing in proper firing order (alpha if numeric & vice versa). Technicians then stamp the employee ID#, Work Order # and date on #1 rod webbing. Stamping this information on the rod webbing helps keep the quality control process order in case of future issues during the manufacturing process.

Step 7: Re-straightening for Rebuilt Crankshafts

The seventh step is industrial crankshaft re-straightening. If the remanufactured crankshaft is deemed un-straight than technicians use an industrial straightening machine on the crankshaft. The straightening machine determines how many dials are out of line. To re-straighten the shaft technicians heat up the crankshaft to 500-600 degrees. Any more than 700 degrees takes the hardness out of the shaft. The strightener process corrects the bent crankshaft to the proper OEM specifications for rebuilt crankshafts.

Step 8: Repeat Magnetic Particle Inspection Process

The eight step in the process is repeating the magnetic particle inspection process if straightening was performed. Anytime metal is being stressed it is imperative to re-inspect for cracks and structural imperfections on the reman crankshaft.

Step 9: Undercutting

The ninth step in the industrial crankshaft re-manufacturing process is undercutting. Technicians undercut the rod or journals to eliminate wear before buildup.

Step 10: Thermal Spraying

The tenth step is the prevention of further buildup via metalizing often called thermal spraying. Thermal spraying has been around for well over 100 years but is still widely known as the best preventative corrosion fighting technique in the world. Thermal spraying is also known for changing the surface of the metallic component and is common with rebuilt crankshafts. Thermal spraying involves protrusion of molten particles onto the heated metallic surface where is bonds and forms a smooth coating interwoven into the structure. There are many different types of thermal spray alloys that can be employed for re-manufactured crankshafts. Typically, boron alloys are used as they very dense, hard and are oxide free. They also prevent against abrasive materials that cause divots, scratches and cracks in addition to preventing surface erosion and corrosion. Thermal spray is an important step some machine shops employ, but not always performed in the industry.

Step 11: Industrial Crankshaft Welding

The welding process for re-manufactured crankshafts is called submerged arc welding. It is a powdered flux plus a weld which combines to produce a more precise weld. The most common flux powder used is called #1 Flux 2245 HD. This powder eliminates the need for technicians to wear weld masking and reduces the amount of dust by-product.

Step 12: Relieve Structural Stress

The twelfth step is to relieve stress upon the entire rebuilt crankshaft structure by heating it up again to 500-600 degrees.

Step 13: Recheck for Straightness

Next step is to check for overall straightness of the re-manufactured crankshaft once again. If the re-manufactured crankshaft is out of alignment then the technician repeats step 7 and re-straighten the structure. Each of the re-manufactured crankshafts is checked multiple times throughout the re-manufacturing process to ensure quality control. If the straightness is not compromised the rebuilt crankshafts can proceed to step thirteen which is crankshaft grinding.

Step 14: Rough Crankshaft Grinding

This is one of the most important steps in the re-manufacturing process of industrial crankshafts. This step involves rough grinding the excess material from the rod or journals and is known as crankshaft grinding. On the rod there are various mains that need to be reground to proper OEM specifications. These rods are spun grind to the next under-size using the pultrusion crankshaft grinding machine. Rod mains are ground inside and outside. Machine shops have the ability to "crankshaft grind" to any size to bring back the crankshaft to standard OEM specifications.

Step 15: Finished Crankshaft Grinding

Next the technician performs a finished crankshaft grinding procedure. The finished crankshaft grinding is a more precise grind which reaches the correct OEM specifications. Before the technician starts the crankshaft grinding they should see what crankshaft bearings are available and start from there. For example, the OEM specification for a Caterpillar 3306 Rod is 2.9987" – 3.0003". Top industrial crankshaft grinding technicians always stop at the high end of the tolerance level. Lastly the technician further refines the crankshaft grinding process in during the micro-polishing process at step eighteen.

Step 16: Shot Peening

The next step is to process the industrial crankshaft in using shot peen machinery. Shot peening adds an additional layer of hardness to the re-manufactured crankshaft.

Step 17: Replace or Re-tighten Counterweights

Step 17 involves replacing the counterweights in proper firing order. Either the new counterweights are installed or the old counterweight bolts are re-tightened and tested.

Step 18: Determine Proper Balance

The machine shop then determines if the proper rotational balance of the re-manufactured crankshafts is achieved. In the engine the crankshaft, pistons and rods all in a constant rotation. The counterweights are designed to offset the weight of the rod and the pistons in the engine. When in motion the kinetic energy and the sum of all forces should be equal to zero on all moving parts. If the re-manufactured crankshaft counterweights are imbalanced it adds additional stress on other components of the engine. The technician should then make sure the internal balance and the external balance of the crankshaft counterweights are properly aligned.

Step 19: Micro-polishing

Then the technician micro-polishes each of the rebuilt crankshafts by hand. To further refine the crankshaft grinding process the machinist makes the most precise fit by micro-polishing the component with a 600 grit emery cloth. Through micro-polishing and industrial crankshaft grinding, the machine shop achieves the recommended Rockwell hardness and Ra finish (Roughness Parameter).

Step 20: Test Reman Crankshaft Rockwell Hardness

Next the technician checks the industry standard hardness. Industry standards crankshaft hardness is 40 on the Rockwell hardness scale. A 45-50 rating is what most reputable machine shops try to employ for all remanufactured crankshafts. When possible it is wise to go beyond industry standards to prevent any future weaknesses within the unit. Typically, hardness can be reduced if the engine is out of oil or the journal is spun incorrectly.

Step 21: Final Quality Control Inspection

Quality control inspects all of the finished reman crankshafts for internal and external mistakes. A typical quality control department uses separate testing and analytical measurement tools from the technicians to ensure accuracy. If the rebuilt crankshaft passes the quality control inspection it goes onto the rust proofing stage.

Step 22: Rustproof Remanufactured Crankshaft

The vast majority of machine shops apply rust proofing to all remanufactured crankshafts using Cosmoline, which is standard rust-proofing for engine parts.

Step 23: Packaging

Lastly the machine shop packs the finished rebuilt crankshaft correctly making sure to using proper boxing and damage proof coverings. It is important to cover the rod journals (varies per crankshaft) with paper & tape in place.

Microfinishing

To achieve the required specifications, automotive manufacturers which design and produce high-volume, low-cost powertrain components, strive toward surpassing stringent emissions and efficiency regulations (see Euro 6c standards for reference) to reduce losses. In motorsport, powertrain developers strive to increase power output by reducing weight, using strong metal alloys, hardening crankshafts, improving balance, reducing friction and vibration as previously described.

To achieve the required specifications, automotive and motorsport powertrain designers and manufacturers adopt a process called microfinishing. Microfinishing (or superfinishing) is an engineering function concerned with metrology and tribology. Microfinishing takes place after the crankshaft grinding process, and is used to improve the geometry of the crankshaft journals from waviness, peaks and lapping caused by the grinding process and establish surface roughness as low as R_a = 0.01 μm if required.

Microfinished crankshafts show improved roundness and cylindricity for each main and pin and thrust journal, and where applicable the oil seal journal. Another important function to a geometrically correct shape is to provide it with a specific surface roughness as per design requirements for optimum lubrication hydrodynamics (essential for crankshafts in engines with stop/start fuel saving technology).

Today, crankshafts used in outboard engines, motorbikes, cars, trucks, busses, marine engines and electric generators and racing engines, are all microfinished for optimum performance. They are designed and manufactured to transfer as much energy to the fly-wheel and drivetrain and absorb as much power from the con-rods, as efficiently as possible for as long as possible.

With this new technology, a light weight, turbocharged 2.0 liter, 4 cylinder diesel engine, (with a low-cost 4 pin, 5 main induction hardened, cast steel, microfinished crankshaft), in a small family car, potentially delivers 180 hp and provides an average fuel consumption of 60 miles per gallon and beyond.

Fatigue Strength

The fatigue strength of crankshafts is usually increased by using a radius at the ends of each main and crankpin bearing. The radius itself reduces the stress in these critical areas, but since the radius in most cases is rolled, this also leaves some compressive residual stress in the surface, which prevents cracks from forming.

Hardening

Most production crankshafts use induction hardened bearing surfaces, since that method gives good results with low costs. It also allows the crankshaft to be reground without re-hardening. But high performance crankshafts, billet crankshafts in particular, tend to use nitridization instead. Nitridization is slower and thereby more costly, and in addition it puts certain demands on the alloying metals in the steel to be able to create stable nitrides. The advantage of nitridization is that it can be done at low temperatures, it produces a very hard surface, and the process leaves some compressive residual stress in the surface, which is good for fatigue properties. The low temperature during treatment is advantageous in that it doesn't have any negative effects on the steel, such as annealing. With crankshafts that operate on roller bearings, the use of carburization tends to be favored due to the high Hertzian contact stresses in such an application. Like nitriding, carburization also leaves some compressive residual stresses in the surface.

Counterweights

Some expensive, high performance crankshafts also use heavy-metal counterweights to make the crankshaft more compact. The heavy-metal used is most often a tungsten alloy but depleted uranium has also been used. A cheaper option is to use lead, but compared with tungsten its density is much lower.

Stress on Crankshafts

The shaft is subjected to various forces but generally needs to be analysed in two positions. Firstly, failure may occur at the position of maximum bending; this may be at the centre of the crank or at either end. In such a condition the failure is due to bending and the pressure in the cylinder is maximal. Second, the crank may fail due to twisting, so the conrod needs to be checked for shear at the position of maximal twisting. The pressure at this position is the maximal pressure, but only a fraction of maximal pressure.

References

- The Bosch Automotive Handbook, 8th Edition, Bentley Publishers, copyright May 2011, ISBN 978-0-8376-1686-5, pp 581–585.

- V.A.W., Hillier (1991). "74: The ignition system". Fundamentals of Motor Vehicle Technology (4th ed.). Stanley Thornes. p. 450. ISBN 0-7487-05317.

- Nunney, Malcolm James (2007) "The Reciprocating Piston Petrol Engine: Gudgeon pins and their location" Light and heavy vehicle technology (4th ed.) Butterworth-Heinemann, Oxford, UK, p. 28, ISBN 978-0-7506-8037-0

- Hillier, Victor Albert Walter and Pittuck, Frank William (1991) "The Petrol Engine: Gudgeon pins" Fundamentals of Motor Vehicle Technology (4th ed.) Stanley Thornes Pub., Cheltenham, England, p. 34 ISBN 0-7487-0531-7

- Sally Ganchy; Sarah Gancher (2009), Islam and Science, Medicine, and Technology, The Rosen Publishing Group, p. 41, ISBN 1-4358-5066-1

- Sally Ganchy, Sarah Gancher (2009), Islam and Science, Medicine, and Technology, The Rosen Publishing

Group, p. 41, ISBN 1-4358-5066-1

- "What's the hardest alloy of aluminum? [Archive] - Practical Machinist - Largest Manufacturing Technology Forum on the Web". www.practicalmachinist.com. Retrieved 2016-02-05.

- "Internal Combustion Engine". The Gale Encyclopedia of Science. Gale Group via HighBeam Research. Retrieved 3 May 2012.

4

An Integrated Study of Ignition System

An ignition system heats an electrode. The electrode is heated to a very high temperature, mainly to ignite a fuel air mixture in this process. Types of ignition systems that have been developed are the ignition magneto ignition system, laser ignition system and inductive discharge ignition. This text provides the reader with an integrated study of ignition system.

Ignition System

An ignition system generates a spark or heats an electrode to a high temperature to ignite a fuel-air mixture in spark ignition internal combustion engines oil-fired and gas-fired boilers, rocket engines, etc. The widest application for spark ignition internal combustion engines is in petrol road vehicles: cars (autos), four-by fours (SUVs), pickups, vans, trucks, buses.

Compression ignition Diesel engines ignite the fuel-air mixture by the heat of compression and do not need a spark. They usually have glowplugs that preheat the combustion chamber to allow starting in cold weather. Other engines may use a flame, or a heated tube, for ignition. While this was common for very early engines it is now rare.

The first electric spark ignition was probably Alessandro Volta's toy electric pistol from the 1780s.

History

Magneto Systems

Magneto ignition coil.

The simplest form of spark ignition is that using a magneto. The engine spins a magnet inside a coil, or, in the earlier designs, a coil inside a fixed magnet, and also operates a contact breaker, interrupting the current and causing the voltage to be increased sufficiently to jump a small gap. The spark plugs are connected directly from the magneto output. Early magnetos had one coil, with the contact breaker (sparking plug) inside the combustion chamber. In about 1902, Bosch introduced a double-coil magneto, with a fixed sparking plug, and the contact breaker outside the cylinder. Magnetos are not used in modern cars, but because they generate their own electricity they are often found on small engines such as those found in mopeds, lawnmowers, snowblowers, chainsaws, etc. where a battery-based electrical system is not present for any combination of necessity, weight, cost, and reliability reasons. They are also used on piston-engined aircraft engines. Although an electrical supply is available, magneto systems are used mainly because of their higher reliability.

Magnetos were used on the small engine's ancestor, the stationary "hit and miss" engine which was used in the early twentieth century, on older gasoline or distillate farm tractors before battery starting and lighting became common, and on aircraft piston engines. Magnetos were used in these engines because their simplicity and self-contained operation was more reliable, and because magnetos weighed less than having a battery and dynamo or alternator.

Aircraft engines usually have dual magnetos to provide redundancy in the event of a failure, and to increase efficiency by thoroughly and quickly burning the fuel air mix from both sides towards the center. The Wright brothers used a magneto invented in 1902 and built for them in 1903 by Dayton, Ohio inventor, Vincent Groby Apple. Some older automobiles had both a magneto system and a battery actuated system (see below) running simultaneously to ensure proper ignition under all conditions with the limited performance each system provided at the time. This gave the benefits of easy starting (from the battery system) with reliable sparking at speed (from the magneto).

Many modern magneto systems (except for small engines) have removed the second (high voltage) coil from the magneto itself and placed it in an external coil assembly similar to the ignition coil described below. In this development, the induced current in the coil in the magneto also flows through the primary of the external coil, generating a high voltage in the secondary as a result. Such a system is referred to as an 'energy transfer system'. Energy transfer systems provide the ultimate in ignition reliability.

Switchable Systems

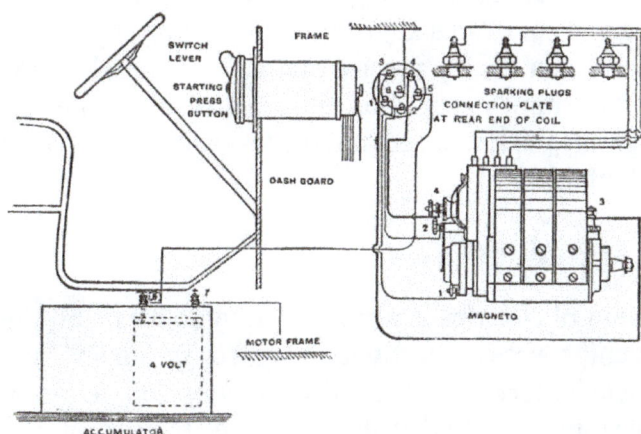

Switchable magneto ignition circuit, with starting battery.

The output of a magneto depends on the speed of the engine, and therefore starting can be problematic. Some magnetos include an impulse system, which spins the magnet quickly at the proper moment, making easier starting at slow cranking speeds. Some engines, such as aircraft but also the Ford Model T, used a system which relied on non rechargeable dry cells, (similar to a large flashlight battery, and which was not maintained by a charging system as on modern automobiles) to start the engine or for starting and running at low speed. The operator would manually switch the ignition over to magneto operation for high speed operation.

To provide high voltage sssfor the spark from the low voltage batteries, a 'tickler' was used, which was essentially a larger version of the once widespread electric buzzer. With this apparatus, the direct current passes through an electromagnetic coil which pulls open a pair of contact points, interrupting the current; the magnetic field collapses, the spring-loaded points close again, the circuit is reestablished, and the cycle repeats rapidly. The rapidly collapsing magnetic field, however, induces a high voltage across the coil which can only relieve itself by arcing across the contact points; while in the case of the buzzer this is a problem as it causes the points to oxidize and/or weld together, in the case of the ignition system this becomes the source of the high voltage to operate the spark plugs.

In this mode of operation, the coil would "buzz" continuously, producing a constant train of sparks. The entire apparatus was known as the 'Model T spark coil' (in contrast to the modern ignition coil which is *only* the actual coil component of the system). Long after the demise of the Model T as transportation they remained a popular self-contained source of high voltage for electrical home experimenters, appearing in articles in magazines such as *Popular Mechanics* and projects for school science fairs as late as the early 1960s. In the UK these devices were commonly known as trembler coils and were popular in cars pre-1910, and also in commercial vehicles with large engines until around 1925 to ease starting.

The Model T (built into the flywheel) differed from modern implementations by not providing high voltage directly at the output; the maximum voltage produced was about 30 volts, and therefore also had to be run through the spark coil to provide high enough voltage for ignition, as described above, although the coil would not "buzz" continuously in this case, only going through one cycle per spark. In either case, the low voltage was switched to the appropriate spark plug by the *'timer'* mounted on the front of the engine. This performed the equivalent function to the modern distributor, although by directing the low voltage, not the high voltage as for the distributor. The timing of the spark was adjustable by rotating this mechanism through a lever mounted on the steering column. As the precise timing of the spark depends on *both* the 'timer' and the trembler contacts within the coil, this is less consistent than the breaker points of the later distributor. However, for the low speed and the low compression of such early engines, this imprecise timing was acceptable.

Battery and Coil-operated Ignition

With the universal adoption of electrical starting for automobiles, and the availability of a large battery to provide a constant source of electricity, magneto systems were abandoned for systems which interrupted current at battery voltage, using an ignition coil (a transformer) to step the voltage up to the needs of the ignition, and a distributor to route the ensuing pulse to the correct spark plug at the correct time.

The first reliable battery operated ignition was developed by the Dayton Engineering Laboratories Co. (Delco) and introduced in the 1910 Cadillac. This ignition was developed by Charles Kettering and was a wonder in its day. It consisted of a single coil, points (the switch), a capacitor and a distributor set up to allocate the spark from the ignition coil timed to the correct cylinder.

The points allow the coil magnetic field to build. When the points open by a cam arrangement, the magnetic field collapses inducing an EMF in the primary that is much larger than the battery voltage and the transformer action produces a large output voltage (20 kV or greater) from the secondary.

The capacitor has two functions. Its main function is to form a parallel resonant circuit with the ignition coil. During resonance, energy is repeatedly transferred to the secondary side until the energy is exhausted. As a result of this resonance the duration of the spark is sustained and so implements a good flame front in the air/fuel mixture. The capacitor's second function and consequent on the first, is to minimise arcing at the contacts at the point of opening by providing an alternative destination for the coil's discharge current. This reduces contact burning and maximizes point life. The Kettering system became the primary ignition system for many years in the automotive industry due to its lower cost, and relative simplicity.

Modern Ignition Systems

The ignition system is typically controlled by a key operated Ignition switch.

Mechanically Timed Ignition

Most four-stroke engines have used a mechanically timed electrical ignition system. The heart of the system is the distributor. The distributor contains a rotating cam driven by the engine's drive, a set of breaker points, a condenser, a rotor and a distributor cap. External to the distributor is the ignition coil, the spark plugs and wires linking the distributor to the spark plugs and ignition coil. (see diagram Below)

Top of distributor cap with wires and terminals

Rotor contacts inside distributor cap

The system is powered by a lead-acid battery, which is charged by the car's electrical system using a dynamo or alternator. The engine operates contact breaker points, which interrupt the current to an induction coil (known as the ignition coil).

The ignition coil consists of two transformer windings — the primary and secondary. These windings share a common magnetic core. An alternating current in the primary induces an alternating magnetic field in the core and hence an alternating current in the secondary. The ignition coil's secondary has more turns than the primary. This is a step-up transformer, which produces a high voltage from the secondary winding. The primary winding is connected to the battery (usually through a current-limiting ballast resistor). Inside the ignition coil one end of each winding is connected together. This common point is taken to the capacitor/contact breaker junction. The other end of the secondary is connected to the rotor. The distributor cap sequences the high voltage to the respective spark plug.

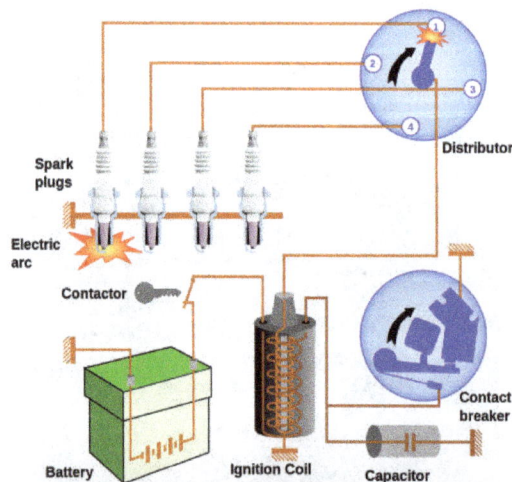

Ignition circuit diagram for mechanically timed ignition

The ignition firing sequence begins with the points (or contact breaker) closed. A steady current flows from the battery, through the current-limiting resistor, through the primary coil, through the closed breaker points and finally back to the battery. This current produces a magnetic field within the coil's core. This magnetic field forms the energy reservoir that will be used to drive the ignition spark.

As the engine turns, the cam inside the distributor rotates. The points ride on the cam so that as a piston reaches the top of the engine's compression cycle, the cam causes the breaker points to open. This breaks the primary winding's circuit and abruptly stops the current through the breaker points. Without the steady current through the points, the magnetic field generated in the coil immediately collapses. This severe rate of change of magnetic flux induces a high voltage in the coil's secondary windings.

At the same time, current exits the coil's primary winding and begins to charge up the capacitor (condenser) that lies across the open breaker points. This capacitor and the coil's primary windings form an oscillating LC circuit. This LC circuit produces a damped, oscillating current which bounces energy between the capacitor's electric field and the ignition coil's magnetic field. The oscillating current in the coil's primary produces an oscillating magnetic field in the coil. This extends the high voltage pulse at the output of the secondary windings. This continues beyond the time of the initial field collapse pulse. The oscillation continues until the circuit's energy is consumed.

The ignition coil's high voltage output is directed to the distributor cap. A turning rotor, located on top of the breaker cam within the distributor cap, sequentially directs the output of the secondary winding to the spark plugs. The high voltage from the coil's secondary (typically 20,000 to 50,000 volts) causes a spark to form across the gap of the spark plug. This, in turn, ignites the compressed air-fuel mixture within the engine. It is the creation of this spark which consumes the energy that was stored in the ignition coil's magnetic field.

The flat twin cylinder 1948 Citroën 2CV used one double ended coil without a distributor, and just contact breakers, in a wasted spark system.

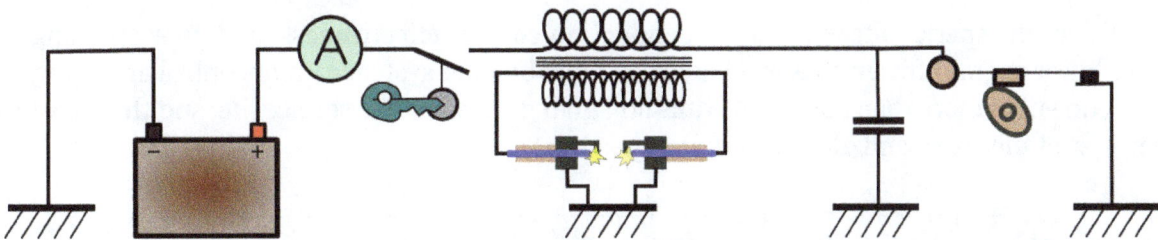

Some two-cylinder motorcycles and motor scooters had two contact points feeding twin coils each connected directly to the spark plug without a distributor; e.g. the BSA Thunderbolt and Triumph Tigress.

High performance engines with eight or more cylinders that operate at high r.p.m. (such as those used in motor racing) demand both a higher rate of spark and a higher spark energy than the simple ignition circuit can provide. This problem is overcome by using either of these adaptations:

- Two complete sets of coils, breakers and condensers can be provided - one set for each half of the engine, which is typically arranged in V-8 or V-12 configuration. Although the two ignition system halves are electrically independent, they typically share a single distributor which in this case contains two breakers driven by the rotating cam, and a rotor with two isolated conducting planes for the two high voltage inputs.

- A single breaker driven by a cam and a return spring is limited in spark rate by the onset of contact bounce or float at high rpm. This limit can be overcome by substituting for the breaker a pair of breakers that are connected electrically in series but spaced on opposite sides of the cam so they are driven out of phase. Each breaker then switches at half the rate of a single breaker and the "dwell" time for current buildup in the coil is maximized since it is shared between the breakers. The Lamborghini V-8 engine has both these adaptations and therefore uses two ignition coils and a single distributor that contains 4 contact breakers.

A distributor-based system is not greatly different from a magneto system except that more separate elements are involved. There are also advantages to this arrangement. For example, the position of the contact breaker points relative to the engine angle can be changed a small amount dynamically, allowing the ignition timing to be automatically advanced with increasing revolutions per minute (RPM) or increased manifold vacuum, giving better efficiency and performance.

However it is necessary to check periodically the maximum opening gap of the breaker(s), using a feeler gauge, since this mechanical adjustment affects the "dwell" time during which the coil charges, and breakers should be re-dressed or replaced when they have become pitted by electric arcing. This system was used almost universally until the late 1970s, when electronic ignition systems started to appear.

Electronic Ignition

The disadvantage of the mechanical system is the use of breaker points to interrupt the low-voltage high-current through the primary winding of the coil; the points are subject to mechanical wear where they ride the cam to open and shut, as well as oxidation and burning at the contact surfaces from the constant sparking. They require regular adjustment to compensate for wear, and the opening of the contact breakers, which is responsible for spark timing, is subject to mechanical variations.

In addition, the spark voltage is also dependent on contact effectiveness, and poor sparking can lead to lower engine efficiency. A mechanical contact breaker system cannot control an average ignition current of more than about 3 A while still giving a reasonable service life, and this may limit the power of the spark and ultimate engine speed.

Example of a basic electronic ignition system.

Electronic ignition (EI) solves these problems. In the initial systems, points were still used but they handled only a low current which was used to control the high primary current through a solid state switching system. Soon, however, even these contact breaker points were replaced by an angular sensor of some kind - either optical, where a vaned rotor breaks a light beam, or more commonly using a Hall effect sensor, which responds to a rotating magnet mounted on the distributor shaft. The sensor output is shaped and processed by suitable circuitry, then used to trigger a switching device such as a thyristor, which switches a large current through the coil.

The first electronic ignition (a cold cathode type) was tested in 1948 by Delco-Remy, while Lucas introduced a transistorized ignition in 1955, which was used on BRM and Coventry Climax Formula One engines in 1962. The aftermarket began offering EI that year, with both the AutoLite Electric Transistor 201 and Tung-Sol EI-4 (thyratron capacitive discharge) being available. Pontiac became the first automaker to offer an optional EI, the breakerless magnetic pulse-triggered Delcotronic, on some 1963 models; it was also available on some Corvettes. The first commercially available all solid-state (SCR) capacitive discharge ignition was manufactured by Hyland Electronics in Canada also in 1963. Ford fitted a Lucas system on the Lotus 25s entered at Indianapolis the next year, ran a fleet test in 1964, and began offering optional EI on some models in 1965. Beginning in 1958, Earl W. Meyer at Chrysler worked on EI, continuing until 1961 and resulting in use of EI on the company's NASCAR hemis in 1963 and 1964.

Prest-O-Lite's CD-65, which relied on capacitance discharge (CD), appeared in 1965, and had "an unprecedented 50,000 mile warranty." (This differs from the non-CD Prest-O-Lite system introduced on AMC products in 1972, and made standard equipment for the 1975 model year.) A similar CD unit was available from Delco in 1966, which was optional on Oldsmobile, Pontiac, and GMC vehicles in the 1967 model year. Also in 1967, Motorola debuted their breakerless CD system. The most famous aftermarket electronic ignition which debuted in 1965, was the Delta Mark 10 capacitive discharge ignition, which was sold assembled or as a kit.

The Fiat Dino is the first production car to come standard with EI in 1968, followed by the Jaguar XJ Series 1 in 1971, Chrysler (after a 1971 trial) in 1973 and by Ford and GM in 1975.

In 1967, Prest-O-Lite made a "Black Box" ignition amplifier, intended to take the load off of the distributor's breaker points during high r.p.m. runs, which was used by Dodge and Plymouth on their factory Super Stock Coronet and Belvedere drag racers. This amplifier was installed on the interior side of the cars' firewall, and had a duct which provided outside air to cool the unit. The rest of the system (distributor and spark plugs) remains as for the mechanical system. The lack of moving parts compared with the mechanical system leads to greater reliability and longer service intervals.

Chrysler introduced breakerless ignition in mid-1971 as an option for its 340 V8 and the 426 Street Hemi. For the 1972 model year, the system became standard on its high-performance engines (the 340 cu in (5.6 l) and the four-barrel carburetor-equipped 400 hp (298 kW) 400 cu in (7 l)) and was an option on its 318 cu in (5.2 l), 360 cu in (5.9 l), two-barrel 400 cu in (6.6 l), and low-performance 440 cu in (7.2 l) . Breakerless ignition was standardised across the model range for 1973.

For older cars, it is usually possible to retrofit an EI system in place of the mechanical one. In some cases, a modern distributor will fit into the older engine with no other modifications, like the H.E.I. distributor made by General Motors, the Hot-Spark electronic ignition conversion kit, and the Chrysler breakerless system.

Coil pack from Honda (one of six).

Other innovations are currently available on various cars. In some models, rather than one central coil, there are individual coils on each spark plug, sometimes known as direct ignition or coil on plug (COP). This allows the coil a longer time to accumulate a charge between sparks, and therefore a higher energy spark. A variation on this has each coil handle two plugs, on cylinders which are 360 degrees out of phase (and therefore reach TDC at the same time); in the four-cycle engine this means that one plug will be sparking during the end of the exhaust stroke while the other fires at the usual time, a so-called "wasted spark" arrangement which has no drawbacks apart from faster spark plug erosion; the paired cylinders are 1/4 and 2/3. Other systems do away with the distributor as a timing apparatus and use a magnetic crank angle sensor mounted on the crankshaft to trigger the ignition at the proper time.

Digital Electronic Ignitions

At the turn of the 21st century digital electronic ignition modules became available for small engines on such applications as chainsaws, string trimmers, leaf blowers, and lawn mowers. This was made possible by low cost, high speed, and small footprint microcontrollers. Digital electronic ignition modules can be designed as either capacitor discharge ignition (CDI) or inductive discharge ignition (IDI) systems. Capacitive discharge digital ignitions store charged energy for the spark in a capacitor within the module that can be released to the spark plug at virtually any time throughout the engine cycle via a control signal from the microprocessor. This allows for greater timing flexibility, and engine performance; especially when designed hand-in-hand with the engine carburetor.

Engine Management

In an Engine Management System (EMS), electronics control fuel delivery and ignition timing. Primary sensors on the system are crankshaft angle (crankshaft or Top Dead Center (TDC) position), airflow into the engine and throttle position. The circuitry determines which cylinder needs fuel and how much, opens the requisite injector to deliver it, then causes a spark at the right moment to burn it. Early EMS systems used an analogue computer to accomplish this, but as embedded systems dropped in price and became fast enough to keep up with the changing inputs at high revolutions, digital systems started to appear.

Some designs using an EMS retain the original ignition coil, distributor and high-tension leads found on cars throughout history. Other systems dispense with the distributor altogether and have individual coils mounted directly atop each spark plug. This removes the need for both distributor and high-tension leads, which reduces maintenance and increases long-term reliability.

Modern EMSs read in data from various sensors about the crankshaft position, intake manifold temperature, intake manifold pressure (or intake air volume), throttle position, fuel mixture via the oxygen sensor, detonation via a knock sensor, and exhaust gas temperature sensors. The EMS then uses the collected data to precisely determine how much fuel to deliver and when and how far to advance the ignition timing. With electronic ignition systems, individual cylinders can have their own individual timing so that timing can be as aggressive as possible per cylinder without fuel detonation. As a result, sophisticated electronic ignition systems can be both more fuel efficient, and produce better performance over their counterparts.

Turbine, Jet and Rocket Engines

Gas turbine engines, including jet engines, have a CDI system using one or more ignitor plugs, which are only used at startup or in case the combustor(s) flame goes out.

Rocket engine ignition systems are especially critical. If prompt ignition does not occur, the combustion chamber can fill with excess fuel and oxidiser and significant overpressure can occur (a "hard start") or even an explosion. Rockets often employ pyrotechnic devices that place flames across the face of the injector plate, or, alternatively, hypergolic propellants that ignite spontaneously on contact with each other. The latter types of engines do away with ignition systems entirely and cannot experience hard starts, but the propellants are highly toxic and corrosive.

Types of Ignition System

Ignition Magneto

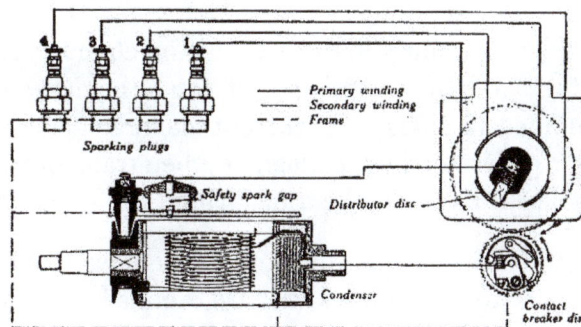

Bosch magneto circuit, 1911

Simple low-tension magneto, for a single-cylinder engine

Armature of a high-tension magneto

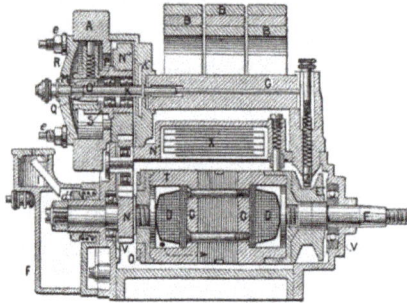

Section through a high-tension magneto, with distributor

An ignition magneto, or high tension magneto, is a magneto that provides current for the ignition system of a spark-ignition engine, such as a petrol engine. It produces pulses of high voltage for the spark plugs. The older term *tension* means *voltage*.

The use of ignition magnetos is now confined mainly to engines where there is no other available electrical supply, for example in lawnmowers and chainsaws. It is also widely used in aviation piston engines even though an electrical supply is usually available. In this case the magneto's self-powered operation is considered to offer increased reliability; in theory the magneto should continue operation as long as the engine is turning.

History

Firing the gap of a spark plug, particularly in the combustion chamber of a high-compression engine, requires a greater voltage (or *higher tension*) than can be achieved by a simple magneto. The *high-tension magneto* combines an alternating current magneto generator and a transformer. A high current at low voltage is generated by the magneto, then transformed to a high voltage (even though this is now a far smaller current) by the transformer.

The first person to develop the idea of a high-tension magneto was Andre Boudeville, but his design omitted a condenser (capacitor); Frederick Richard Simms in partnership with Robert Bosch were the first to develop a practical high-tension magneto.

Magneto ignition was introduced on the 1899 Daimler Phönix. This was followed by Benz, Mors, Turcat-Mery, and Nesseldorf, and soon was used on most cars up until about 1918 in both low voltage (voltage for secondary coils to fire the spark plugs) and high voltage magnetos (to fire the spark plug directly, similar to coil ignitions, introduced by Bosch in 1903).

Operation

In the type known as a *shuttle magneto*, the engine rotates a coil of wire between the poles of a magnet. In the *inductor magneto*, the magnet is rotated and the coil remains stationary.

As the magnet moves with respect to the coil, the magnetic flux linkage of the coil changes. This induces an EMF in the coil, which in turn causes a current to flow. One or more times per revolution, just as the magnet pole moves away from the coil and the magnetic flux begins to decrease, a cam opens the contact breaker and interrupts the current. This causes the electromagnetic field in the primary coil to collapse rapidly. As the field collapses rapidly there is a large voltage induced (as described by Faraday's Law) across the primary coil.

As the points begin to open, point spacing is initially such that the voltage across the primary coil would arc across the points. A capacitor is placed across the points which absorbs the energy stored in the leakage inductance of the primary coil, and slows the rise time of the primary winding voltage to allow the points to open fully. The capacitor's function is similar to that of a snubber as found in a flyback converter.

A second coil, with many more turns than the primary, is wound on the same iron core to form an electrical transformer. The ratio of turns in the secondary winding to the number of turns in the primary winding, is called the *turns ratio*. Voltage across the primary coil results in a proportional voltage being induced across the secondary winding of the coil. The turns ratio between the primary and secondary coil is selected so that the voltage across the secondary reaches a very high value, enough to arc across the gap of the spark plug. As the voltage of the primary winding rises to several hundred volts, the voltage on the secondary winding rises to several tens of thousands of volts, since the secondary winding typically has 100 times as many turns as the primary winding.

The capacitor and the coil together form a resonant circuit which allows the energy to oscillate from the capacitor to the coil and back again. Due to the inevitable losses in the system, this oscillation decays fairly rapidly. This dissipates the energy that was stored in the condenser in time for the next closure of the points, leaving the condenser discharged and ready to repeat the cycle.

On more advanced magnetos the cam ring can be rotated by an external linkage to alter the ignition timing.

In a modern installation, the magneto only has a single low tension winding which is connected to an external ignition coil which not only has a low tension winding, but also a secondary winding of many thousands of turns to deliver the high voltage required for the spark plug(s). Such a system is known as an "energy transfer" ignition system. Initially this was done because it was easier to provide good insulation for the secondary winding of an external coil than it was in a coil buried in the construction of the magneto (early magnetos had the coil assembly externally to the rotating parts to make them easier to insulate—at the expense of efficiency). In more modern times, insulation materials have improved to the point where constructing self-contained magnetos is relatively easy, but energy transfer systems are still used where the ultimate in reliability is required such as in aviation engines.

Aviation

Because it requires no battery or other source of electrical energy, the magneto is a compact and reliable self-contained ignition system, which is why it remains in use in many general aviation applications.

Since the beginning of World War I in 1914, magneto-equipped aircraft engines have typically been *dual-plugged*, whereby each cylinder has two spark plugs, with each plug having a separate magneto system. Dual plugs provide both redundancy should a magneto fail, and better engine performance (through enhanced combustion). Twin sparks provide two flame fronts within the cylinder, these two flame fronts decreasing the time needed for the fuel charge to burn, thereby burning more of the fuel at a lower temperature and pressure. As the pressure within a cylinder increases, the temperature rises; and if there is only a single plug, the unburnt fuel away from the original flame front can self-ignite, producing a separate unsynchronized flame front. This leads to a rapid rise in cylinder pressure, producing engine "knock". Higher octane fuel delays the time required for auto-ignition at a given temperature and pressure, reducing knock; so by burning the fuel charge faster, two flame fronts can decrease an engine's octane requirement. As the size of the combustion chamber determines the time to burn the fuel charge, dual ignition was especially important for the large-bore aircraft engines around World War II.

Impulse Coupling

Because the magneto has low voltage output at low speed, starting an engine is more difficult. Therefore, some magnetos have an impulse coupling, a springlike mechanical linkage between the engine and magneto drive shaft which "winds up" and "lets go" at the proper moment for spinning the magneto shaft. The impulse coupling uses a spring, a hub cam with flyweights, and a shell. The hub of the magneto rotates while the drive shaft is held stationary, and the spring tension builds up. When the magneto is supposed to fire, the flyweights are released by the action of the body contacting the trigger ramp. This allows the spring to unwind giving the rotating magnet a rapid rotation and letting the magneto spin at such a speed to produce a spark.

Automobile

Some aviation engines as well as some early luxury cars have had dual-plugged systems with one set of plugs fired by a magneto, and the other set wired to a coil, dynamo, and battery circuit. This was often done to ease engine starting, as larger engines may be too difficult to crank at sufficient speed to operate a magneto, even with an impulse coupling. As the reliability of battery ignition systems improved, the magneto fell out of favour for general automotive use, but may still be found in sport or racing engines.

Inductive Discharge Ignition

Inductive discharge ignition systems were developed in the 19th century as a means to ignite the air-fuel mixture in the combustion chamber of internal combustion engines. The first versions were low tension coils, then low-tension and in turn high-tension magnetos, which were offered as a more effective alternative to the older-design hot-tube ignitors that had been utilized earlier on hot tube engines. With the advent of small stationary engines; and with the development of the automobile, engine-driven tractors, and engine-driven trucks; first the magneto and later the distributor-type systems were utilized as part of an efficient and reliable engine ignition system on commercially available motorized equipment. These systems were in widespread use on all cars and trucks through the 1960s. Manufacturers such as Ford, General Motors, Chrysler, Citroen, Mercedes, John Deere, International Harvester, and many others incorporated them into their products. The inductive discharge system is still extensively used today.

Faraday's Law

The inductive-discharge ignition system operates according to the rules of electromagnetism described by Faraday's Law of Induction which states that the induction of electromotive force (emf) in any closed circuit is equal to the time rate of change of the magnetic flux through the circuit. In other words, the emf generated is proportional to the rate of change of the magnetic flux. More simply stated, an electric field is induced in any system in which a magnetic field is changing with time. The change could be changes in direction of force or strength. The effects described by this law are those by which generators, motors, alternators, and transformers function. There are two main concepts to be taken from Faraday's Law that apply to the design of inductive discharge ignitions. One is that moving a wire through a magnetic field will induce an electric voltage and current in the wire, aka electromagnetic induction. The second is that current moving in a wire will induce a magnetic field around the wire.

Magnetos

A magneto is one of the electromechanical devices invented for the purpose of ignition with gasoline internal combustion engines. A magneto at its most basic is a simple magnet that moves next to a wire, or sometimes a wire moves next to a magnet. As they move in relation to each other, the changes in direction of magnetic force induce an electric current in the wire. Usually the wire (called a primary wire) is very long, and looped around an iron magnetic core that more or less channels the magnetic field through the loop of wire. As the current flows, the wire loops develop their own magnetic field, which takes a certain amount of energy to form. The magnetic field is a type of potential energy. There is usually some sort of device that opens and closes the circuit called a contact breaker, points or an ignitor. As the points or ignitor open, the current ceases flowing, and the magnetic field collapses. The energy stored in the magnetic field is released in the form of increased electric voltage in the wire. This voltage jumps across the gap of either the ignitor or a spark plug located in the combustion chamber and ignites the air-fuel mixture to do work.

Some magnetos have a second coil of wire located next to the first, called a secondary coil. This coil is usually much longer than the primary loop, accomplished by many more loops around the magnetic core. As the magnetic field is built, it induces a current in the secondary coil as well. When the contact breaker opens the circuit, the magnetic field collapses, causing a high electric voltage in the primary and secondary coils. However, due to the greater number of turns of the secondary coil, the voltage is much higher, causing a larger spark at the ignitor or spark plug, meaning more assured ignition.

Due to their reliability, magnetos are used as ignition systems on aircraft. They are also used on machinery that do not have a separate electric supply or battery. They are also used on drag race cars because they offer a weight advantage over systems that utilize a distributor and battery.

Distributor Ignition Systems

As ignition technology developed, engineers realized that a functional ignition system could be designed that dispensed with the magnets altogether. By applying a current to a primary wire loop wrapped around an iron magnetic core, a magnetic field would be generated in the primary loop without the magnets. This magnetic field would induce a current in an adjacent longer secondary loop of wire. By opening the circuit in the primary loop, the collapsing magnetic field would cause a voltage to be induced in the secondary loop. This high voltage was carried or "distributed" by a distributor to each of the multiple spark plugs in a gasoline car or truck engine.

The most familiar version of this kind of system was invented by Charles F. Kettering in about 1909 and was known by some as the Delco ignition system. Later patent applications to the US Patent Office make reference to the "Kettering ignition system". This type of ignition system was used on automobiles, trucks, lawn mowers, tractors, chainsaws, and other gasoline-powered machinery with great success for many decades until the development of capacitive-discharge ignition systems.

Laser Ignition

Laser ignition is an alternative method for igniting mixtures of fuel and oxidiser. The phase of the mixture can be gaseous or liquid. The method is based on a laser igniton devices that produce short but powerful flashes regardless of the pressure in the combustion chamber. Usually, high voltage spark plugs are good enough for automotive use, as the typical compression ratio of an otto cycle internal combustion engine is around 10:1 and in some rare cases reach 14:1. However, fuels such as natural gas or methanol can withstand high compression without self ignition. This allows higher compression ratios, because it is economically reasonable, as the fuel efficiency of such engines is high. Using high compression ratio and high pressure requires special spark plugs that are expensive and their electrodes still wear out. Thus, even expensive laser ignition systems could be economical, because they would last longer.

Further Applications of Laser Ignition

Laser ignition is considered as a potential ignition system for non-hypergolic liquid rocket engines and reaction control systems which need an ignition system. Conventional ignition technologies like torch igniters are more complex in sequencing and need additional components like propellant feed lines and valves. Therefore, they are heavy compared to a laser ignition system. Pyrotechnical devices allow only one ignition per unit and imply increased launch pad precautions as they are made of explosives.

Hot-wiring

Hot-wiring is the process of bypassing an automobile's ignition interlock and thus starting it without the key. A vehicle owner who has lost their vehicle key may implement this process.

Methods

Hot-wiring generally involves connecting the two wires which complete the circuit when the key is in the "on" position (turning on the fuel pump and other necessary components), then touching the wire that connects to the starter. The specific method of hot-wiring a vehicle is dependent on the particular vehicle's electrical ignition system. Remote start units access the same wires as conventional ignition methods. Listings of wire colors and locations and ignition system schematics may sometimes be found in Internet databases.

Older vehicles, especially ones made before 1986, which have a carbureted engine and a single ignition coil and distributor, can be hot-wired from the engine bay. Using standard lock picking to start a car is now usually ineffective, since most cars now use electronic chip or transponder verification.

Thieves lacking the basic mechanical skills and knowledge of automotive electrical systems sometimes simply use physical force to bypass the ignition lock, smashing the key mechanism to reveal the rotation switch, which is operated by the key's tumbler.

References

- Munday, Frank (2006). Custom Auto Electrickery: How to Work with and Understand Auto Electrical Systems. MBI Publishing Company. p. 59. ISBN 0-949398-35-7.

- Emanuel, Dave (1996). Small-block Chevy performance: modifications and dyno-tested combinations for high performance street and racing use. Penguin. p. 122. ISBN 1-55788-253-3.

- Huzel, Dieter K. (1992-01-01). Modern Engineering for Design of Liquid-Propellant Rocket Engines. AIAA. ISBN 9781600864001.

- Manfletti, Chiara (2014-01-01). "Laser Ignition of an Experimental Cryogenic Reaction and Control Thruster: Ignition Energies". Journal of Propulsion and Power. 30 (4): 952–961. doi:10.2514/1.B35115. ISSN 0748-4658.

- "New way to get that vital spark - University of Liverpool". Liv.ac.uk. 2008-10-31. Archived from the original on 2014-01-10. Retrieved 2014-02-01.

5

Important Concepts and Principles of Internal Combustion Engine

This chapter elucidates the important concepts and principles of the internal combustion engine. The important topics explained in this chapter are forced induction, manifold vacuum, consumption map and brake specific fuel consumption. The section strategically encompasses all the important concepts and helps the reader develop a better understanding on the principles of the internal combustion engine.

Forced Induction

Forced induction is the process of delivering compressed air to the intake of an internal combustion engine. A forced induction engine uses a gas compressor to increase the pressure, temperature and density of the air. An engine without forced induction is considered a naturally aspirated engine.

Introduction

Forced induction is used in the automotive and aviation industry to increase engine power and efficiency. A forced induction engine is essentially two compressors in series. The compression stroke of the engine is the main compression that every engine has. An additional compressor feeding into the intake of the engine causes forced induction of air. A compressor feeding pressure into another greatly increases the total compression ratio of the entire system. This intake pressure is called boost. This particularly helps aviation engines, as they need to operate at higher altitudes with lower air densities.

Higher compression engines have the benefit of maximizing the amount of useful energy evolved per unit of fuel. Therefore, the thermal efficiency of the engine is increased in accordance with the vapour power cycle analysis of the second law of thermodynamics. The reason all engines are not higher compression is because for any given octane, the fuel will prematurely detonate with a higher than normal compression ratio. This is called preignition, detonation or knock and can cause severe engine damage. High compression on a naturally aspirated engine can reach the detonation threshold fairly easily. However, a forced induction engine can have a higher total compression without detonation because the air charge can be cooled after the first stage of compression, using an intercooler.

One of the primary concerns in internal combustion emissions is a factor called the NOx fraction, or the amount of nitrogen/oxygen compounds the engine produces. This level is government regulated for emissions as commonly seen at inspection stations. High compression causes high combustion temperatures. High combustion temperatures lead to higher NOx emissions, thus forced induction can give higher NOx fractions.

Types of Compressors

Two commonly used forced-induction compressors are turbochargers and superchargers. A turbocharger is a centrifugal compressor driven by the flow of exhaust gasses. Superchargers use various different types of compressors but are all powered directly by the rotation of the engine, usually through a belt drive. The compressor can be centrifugal or a Roots-type for positive displacement compression. An example of an internal compressor is a screw-type supercharger or a piston compressor.

Turbochargers

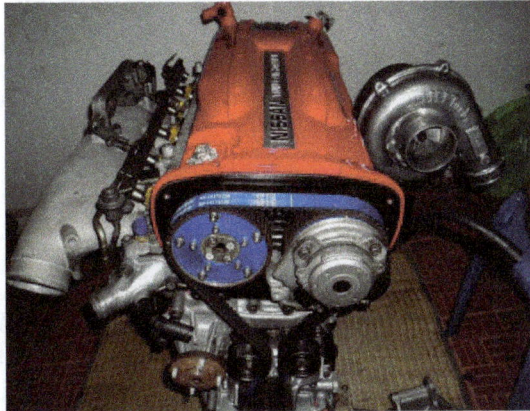

A turbocharged Nissan RB engine.

A turbocharger relies on the volume and velocity of exhaust gases to spin (spool) the turbine wheel, which is connected to the compressor wheel via a common shaft. The boost pressure made can be regulated by a system of release valves and electronic controllers. The chief benefit of a turbocharger is that it consumes less power from the engine than a supercharger; the main drawback is that engine response suffers greatly because it takes time for the turbocharger to come up to speed (spool up). This delay in power delivery is referred to as turbo lag. Any given turbo design is inherently one of compromise; a smaller turbo will spool quickly and deliver full boost pressure at low engine speeds, but boost pressure will suffer at high engine RPM. A larger turbo, on the other hand, will provide improved high-rev performance at the expense of low-end response. Other common design issues include limited turbine lifespan, due to the high exhaust temperatures it must withstand, and the restrictive effect the turbine has upon exhaust flow.

Superchargers

Superchargers have almost no lag time to build pressure because the compressor is always spinning proportionally to the engine speed. They are not as common as turbochargers because they use the torque produced from the engine to operate. This results in some loss in power and efficiency. A Roots-type supercharger uses paddles on two rotating drums to push air into the intake. Because it is a positive displacement device, this compressor has the advantage of producing the same pressure ratio at any engine speed. A screw-type supercharger is also a positive displacement device, like a Roots-type supercharger. Screw-type superchargers are more complex to manufacture than Roots-type superchargers, but are more efficient to operate, producing cooler air output. A centrifugal-type supercharger is not a positive displacement device and will usually have better thermal efficiency than a Roots-type supercharger. Centrifugal superchargers are also more compact and easier to use with an intercooler.

A Roots-type supercharger on a Nissan VQ engine.

Intercooling

An unavoidable side-effect of forced induction is that compressing air raises its temperature. As a result, the charge density is reduced and the cylinders receive less air than the system's boost pressure prescribes. The risk of detonation, or "knock", greatly increases. These drawbacks are countered by charge-air cooling, which passes the air leaving the turbocharger or supercharger through a heat exchanger typically called an intercooler. This is done by cooling the charge air with an ambient flow of either air (air-air intercooler) or liquid (liquid-to-air intercooler). The charge air density is increased and the temperature is reduced. In this way an intercooler can greatly increase the ability to run higher absolute compression ratios and take full advantage of using compressors in series. The only drawbacks of intercooling are the intercooler's size (typically close to the size of a radiator), and the associated plumbing and piping.

Water Injection

Water injection is another effective means of cooling the charge air to prevent detonation. Methanol is mixed with the water to prevent freezing and to act as a slower-burning fuel. Water injection, unlike nitrous oxide or forced induction, doesn't add much power to the engine by itself, but allows more power to be safely added. It works by being sprayed into the compressed air charge. The water absorbs heat as it evaporates to cool the charge and lower combustion temperatures. The alcohol is also a fuel in the charge which burns slower and cooler than gasoline. Due to the lower intake temperatures and denser air charge, more boost pressure and timing advance can be safely added without using higher octane fuel. It is most often used in racing applications, however it was also shown to be practical for extended use.

Diesel Engines

Diesel engines do not have preignition problems because fuel is injected at the end of the compression stroke, therefore higher compression is used. Most modern diesel engines use a turbocharger. This is because the exhaust from a diesel is exceptionally strong making it excellent for powering a turbo. The range of engine speed is narrower, allowing for a single turbo to fully power the entire engine range. Turbochargers can also achieve higher boost pressure than superchargers, which is necessary for most diesels.

Design Considerations

The design of gasoline engines and the compression ratio impact the maximum possible boost. To obtain more power from higher boost levels and maintain reliability, many engine components have to be replaced or upgraded from that of naturally aspirated powertrains. Design considerations include the fuel pump, fuel injectors, pistons, connecting rods, crankshafts, valves, head-gasket, and head bolts. The maximum possible boost depends on the fuel's octane rating and the inherent tendency of any particular engine toward detonation. Premium gasoline or racing gasoline can be used to prevent detonation within reasonable limits. Ethanol, methanol, liquefied petroleum gas (LPG) and compressed natural gas (CNG) allow higher boost than gasoline, because of their higher resistance to autoignition (lower tendency to knock). Diesel engines can also tolerate much higher levels of boost pressure than Otto cycle engines, because only air is being compressed during the compression phase, and fuel is injected later, removing the knocking issue entirely.

Motorcycles

Unique design considerations for motorcycles include tractable power delivery; and packaging for heat removal, space conservation, and desired center of gravity.

Manifold Vacuum

Manifold vacuum, or engine vacuum in an internal combustion engine is the difference in air pressure between the engine's intake manifold and Earth's atmosphere.

Manifold vacuum is an effect of a piston's movement on the induction stroke and the choked flow through a throttle in the intake manifold of an engine. It is a measure of the amount of restriction of airflow through the engine, and hence of the unused power capacity in the engine. In some engines, the manifold vacuum is also used as an auxiliary power source to drive engine accessories and for the crankcase ventilation system.

Manifold vacuum should not be confused with venturi vacuum, which is an effect exploited in carburetors to establish a pressure difference roughly proportional to mass airflow and to maintain a somewhat constant air/fuel ratio. It is also used in light airplanes to provide airflow for pneumatic gyroscopic instruments.

Overview

The rate of airflow through an internal combustion engine is an important factor determining the amount of power the engine generates. Most gasoline engines are controlled by limiting that flow with a throttle that restricts intake airflow, while a diesel engine is controlled by the amount of fuel supplied to the cylinder, and so has no "throttle" as such. Manifold vacuum is present in all naturally aspirated engines that use throttles (including carbureted and fuel injected gasoline engines using the Otto cycle or the two-stroke cycle; diesel engines do not have throttle plates).

The mass flow through the engine is determined by the rotation rate of the engine, multiplied by the displacement of the engine, and the density of the intake stream in the intake manifold. In most applications the rotation rate is set by the application (engine speed -RPM- in a vehicle or machinery speed in other applications). The displacement is dependent on the engine geometry, which is generally not adjustable while the engine is in use (although a handful of models do have this feature, see variable displacement). Restricting the input flow reduces the density (and hence pressure) in the intake manifold, reducing the amount of power produced. It is also a major source of engine drag (see engine braking), as the engine must pump material from the low-pressure intake manifold into the exhaust manifold (at ambient atmospheric pressure).

When the throttle is opened (in a car, the accelerator pedal is depressed), ambient air is free to fill the intake manifold, increasing the pressure (filling the vacuum). A carburetor or fuel injection system adds fuel to the airflow in the correct proportion, providing energy to the engine. When the throttle is opened all the way, the engine's air induction system is exposed to full atmospheric pressure, and maximum airflow through the engine is achieved. In a naturally aspirated engine, output power is limited by the ambient barometric pressure. Superchargers and turbochargers boost manifold pressure above atmospheric pressure.

Modern Developments

Modern engines use a MaP sensor to measure air pressure in the intake manifold. Manifold Pressure is one of a multitude of parameters used by the engine control unit (ECU) to optimize engine operation.

In the past, this sensor was sometimes referred to as a Manifold Absolute Pressure (MAP) sensor. This terminology is considered obsolete because only absolute pressure is meaningful in an engine-control context and it is unnecessary (and potentially confusing) to specify "absolute". This is especially true in forced induction(turbocharged or supercharged) engines, whose manifold pressures are quite independent of ambient pressure and may be either greater or less than ambient; and aircraft engines, which operate at a wide variety of ambient pressures.

Motivated by government regulations mandating reduction of fuel consumption (in the USA) or reduction of carbon dioxide emissions, (in Europe) passenger cars and light trucks have been fitted with a variety of technologies (downsized engines; lockup, multi-ratio and overdrive transmissions; variable valve timing, boost, Diesel engines, et al.) which render manifold vacuum inadequate or unavailable. Electric vacuum pumps are now commonly used for powering pneumatic accessories.

Manifold Vacuum Vs. Venturi Vacuum

Manifold vacuum is caused by a different phenomenon than venturi vacuum, which is present inside carburetors. Venturi vacuum is caused by the venturi effect which, for fixed ambient conditions (air density and temperature), depends on the total mass flow through the carburetor. In engines that use carburetors, the venturi vacuum is approximately proportional to the total mass flow through the engine (and hence the total power output). As ambient pressure (altitude, weather) or temperature change, the carburetor may need to be adjusted to maintain this relationship.

Manifold pressure may also be "ported". Porting is selecting a location for the pressure tap within the throttle plate's range of motion. Depending on throttle position, a ported pressure tap may be either upstream or downstream of the throttle. As the throttle position changes, a "ported" pressure tap is selectively connected to either manifold pressure or ambient pressure. Antique (pre-OBD II) engines often used ported manifold pressure taps for ignition distributors and emission-control components.

Manifold Vacuum in Cars

Most automobiles use four-stroke Otto cycle engines with multiple cylinders attached to a single inlet manifold. During the induction stroke, the piston descends in the cylinder and the intake valve is open. As the piston descends it effectively increases the volume in the cylinder above it, setting up low pressure. Atmospheric pressure pushes air through the manifold and carburetor or fuel injection system, where it is mixed with fuel. Because multiple cylinders operate at different times in the engine cycle, there is almost constant pressure difference through the inlet manifold from carburetor to engine.

To control the amount of fuel/air mix entering the engine, a simple butterfly valve (the throttle) is generally fitted at the start of the intake manifold (just below the carburetor in carbureted engines). The butterfly valve is simply a circular disc fitted on a spindle, fitting inside the pipe work. It is connected to the accelerator pedal of the car, and is set to be fully open when the pedal is fully depressed and fully closed when the pedal is released. The butterfly valve often contains a small "idle cutout", a hole that allows small amounts of fuel/air mixture into the engine even when the valve is fully closed, or the carburetor has a separate air bypass with its own idle jet.

If the engine is operating under light or no load and low or closed throttle, there is high manifold vacuum. As the throttle is opened, the engine speed increases rapidly. The engine speed is limited only by the amount of fuel/air mixture that is available in the manifold. Under full throttle and light load, other effects (such as valve float, turbulence in the cylinders, or ignition timing) limit engine speed so that the manifold pressure can increase—but in practice, parasitic drag on the internal walls of the manifold, plus the restrictive nature of the venturi at the heart of the carburetor, means that a low pressure will always be set up as the engine's internal volume exceeds the amount of the air the manifold is capable of delivering.

If the engine is operating under heavy load at wide throttle openings (such as accelerating from a stop or pulling the car up a hill) then engine speed is limited by the load and minimal vacuum will be created. Engine speed is low but the butterfly valve is fully open. Since the pistons are descending more slowly than under no load, the pressure differences are less marked and parasitic drag in the induction system is negligible. The engine pulls air into the cylinders at the full ambient pressure.

More vacuum is created in some situations. On deceleration or when descending a hill, the throttle will be closed and a low gear selected to control speed. The engine will be rotating fast because the road wheels and transmission are moving quickly, but the butterfly valve will be fully closed. The flow of air through the engine is strongly restricted by the throttle, producing a strong vacuum on the engine side of the butterfly valve which will tend to limit the speed of the engine. This phenomenon, known as engine braking, is used to prevent acceleration or even to slow down with minimal or no brake usage (as when descending a long or steep hill). This vacuum braking should not be confused with compression braking (aka a "Jake brake"), or with exhaust braking, which are often used on large diesel trucks. Such devices are necessary for engine braking with a diesel as they lack a throttle to restrict the air flow enough to create sufficient vacuum to brake a vehicle.

Uses of Manifold Vacuum

Autovac fuel lifters. On both buses the red Autovac tank can be seen above and behind the left front wheel.

This low (or negative) pressure can be put to uses. A pressure gauge measuring the manifold pressure can be fitted to give the driver an indication of how hard the engine is working and it can be used to achieve maximum momentary fuel efficiency by adjusting driving habits: minimizing manifold vacuum increases momentary efficiency. A weak manifold vacuum under closed-throttle conditions shows that the butterfly valve or internal components of the engine (valves or piston rings) are worn, preventing good pumping action by the engine and reducing overall efficiency.

Vacuum is often used to drive auxiliary systems on the vehicle. Vacuum-assist brake servos, for example, use atmospheric pressure pressing against the engine manifold vacuum to increase pressure on the brakes. Since braking is nearly always accompanied by the closing of the throttle and associated high manifold vacuum, this system is simple and almost foolproof. Vacuum tanks were installed on trailers to control their integrated braking systems.

Prior to the introduction of Federal Motor Vehicle Safety Standards in the USA by the National Traffic and Motor Vehicle Safety Act of 1966, it was common to use manifold vacuum to drive windscreen wipers with a pneumatic motor. This system was cheap & simple but resulted in the comical (and unsafe) effect of wipers which operate at full speed while the engine idles, operate around half speed while cruising, and stop altogether when the driver depresses the pedal fully. Vehicle HVAC systems also used manifold vacuum to drive actuators controlling airflow and temperature.

Another obsolete accessory is the "Autovac" fuel lifter which uses vacuum to raise fuel from the main tank to a small auxiliary tank, from which it flows by gravity to the carburetor. This eliminated the fuel pump which, in early cars, was an unreliable item.

Manifold Vacuum in Diesel Engines

Many diesel engines do not have butterfly valve throttles. The manifold is connected directly to the air intake and the only suction created is that caused by the descending piston with no venturi to increase it, and the engine power is controlled by varying the amount of fuel that is injected into the cylinder by a fuel injection system. This assists in making diesels much more efficient than petrol engines.

If vacuum is required (vehicles that can be fitted with both petrol and diesel engines often have systems requiring it), a butterfly valve connected to the throttle can be fitted to the manifold. This reduces efficiency and is still not as effective as it is not connected to a venturi. Since low-pressure is only created on the overrun (such as when descending hills with a closed throttle), not over a wide range of situations as in a petrol engine, a vacuum tank is fitted.

Most diesel engines now have a separate vacuum pump ("exhauster") fitted to provide vacuum at all times, at all engine speeds.

Many new BMW petrol engines do not use a throttle in normal running, but instead use "Valve-tronic" variable-lift intake valves to control the amount of air entering the engine. Like a diesel engine, manifold vacuum is practically non-existent in these engines and a different source must be utilised to power the brake servo.

Internal Combustion Engine Cooling

Internal combustion engine cooling uses either air or a liquid to remove the waste heat from an internal combustion engine. For small or special purpose engines, air cooling makes for a light-weight and relatively simple system. The more complex circulating liquid-cooled engines also ulti-mately reject waste heat to the air, but circulating liquid improves heat transfer from internal parts of the engine. Engines for watercraft may use open-loop cooling, but air and surface vehicles must recirculate a fixed volume of liquid.

Overview

Heat engines generate mechanical power by extracting energy from heat flows, much as a water wheel extracts mechanical power from a flow of mass falling through a distance. Engines are ineffi-cient, so more heat energy enters the engine than comes out as mechanical power; the difference is waste heat which must be removed. Internal combustion engines remove waste heat through cool intake air, hot exhaust gases, and explicit engine cooling.

Engines with higher efficiency have more energy leave as mechanical motion and less as waste heat. Some waste heat is essential: it guides heat through the engine, much as a water wheel works only if there is some exit velocity (energy) in the waste water to carry it away and make room for more water. Thus, all heat engines need cooling to operate.

Cooling is also needed because high temperatures damage engine materials and lubricants. Cool-ing becomes more important in when the climate becomes very hot. Internal-combustion engines burn fuel hotter than the melting temperature of engine materials, and hot enough to set fire to lubricants. Engine cooling removes energy fast enough to keep temperatures low so the engine can survive.

Some high-efficiency engines run without explicit cooling and with only incidental heat loss, a design called adiabatic. Such engines can achieve high efficiency but compromise power output, duty cycle, engine weight, durability, and emissions.

Basic Principles

Most internal combustion engines are fluid cooled using either air (a gaseous fluid) or a liquid coolant run through a heat exchanger (radiator) cooled by air. Marine engines and some stationary engines have ready access to a large volume of water at a suitable temperature. The water may be used directly to cool the engine, but often has sediment, which can clog coolant passages, or chemicals, such as salt, that can chemically damage the engine. Thus, engine coolant may be run through a heat exchanger that is cooled by the body of water.

Most liquid-cooled engines use a mixture of water and chemicals such as antifreeze and rust inhibitors. The industry term for the antifreeze mixture is *engine coolant*. Some antifreezes use no water at all, instead using a liquid with different properties, such as propylene glycol or a combination of propylene glycol and ethylene glycol. Most "air-cooled" engines use some liquid oil cooling, to maintain acceptable temperatures for both critical engine parts and the oil itself. Most "liquid-cooled" engines use some air cooling, with the intake stroke of air cooling the combustion chamber. An exception is Wankel engines, where some parts of the combustion chamber are never cooled by intake, requiring extra effort for successful operation.

There are many demands on a cooling system. One key requirement is to adequately serve the entire engine, as the whole engine fails if just one part overheats. Therefore, it is vital that the cooling system keep *all* parts at suitably low temperatures. Liquid-cooled engines are able to vary the size of their passageways through the engine block so that coolant flow may be tailored to the needs of each area. Locations with either high peak temperatures (narrow islands around the combustion chamber) or high heat flow (around exhaust ports) may require generous cooling. This reduces the occurrence of hot spots, which are more difficult to avoid with air cooling. Air-cooled engines may also vary their cooling capacity by using more closely spaced cooling fins in that area, but this can make their manufacture difficult and expensive.

Only the fixed parts of the engine, such as the block and head, are cooled directly by the main coolant system. Moving parts such as the pistons, and to a lesser extent the crank and rods, must rely on the lubrication oil as a coolant, or to a very limited amount of conduction into the block and thence the main coolant. High performance engines frequently have additional oil, beyond the amount needed for lubrication, sprayed upwards onto the bottom of the piston just for extra cooling. Air-cooled motorcycles often rely heavily on oil-cooling in addition to air-cooling of the cylinder barrels.

Liquid-cooled engines usually have a circulation pump. The first engines relied on thermo-syphon cooling alone, where hot coolant left the top of the engine block and passed to the radiator, where it was cooled before returning to the bottom of the engine. Circulation was powered by convection alone.

Other demands include cost, weight, reliability, and durability of the cooling system itself.

Conductive heat transfer is proportional to the temperature difference between materials. If engine metal is at 250 °C and the air is at 20 °C, then there is a 230 °C temperature difference for cooling. An air-cooled engine uses all of this difference. In contrast, a liquid-cooled engine might dump heat from the engine to a liquid, heating the liquid to 135 °C (Water's standard boiling point of 100 °C can be exceeded as the cooling system is both pressurised, and uses a mixture with antifreeze) which is then cooled with 20 °C air. In each step, the liquid-cooled engine has half the temperature difference and so at first appears to need twice the cooling area.

However, properties of the coolant (water, oil, or air) also affect cooling. As example, comparing water and oil as coolants, one gram of oil can absorb about 55% of the heat for the same rise in temperature (called the specific heat capacity). Oil has about 90% the density of water, so a given volume of oil can absorb only about 50% of the energy of the same volume of water. The thermal conductivity of water is about 4 times that of oil, which can aid heat transfer. The viscosity of oil can be ten times greater than water, increasing the energy required to pump oil for cooling, and reducing the net power output of the engine.

Comparing air and water, air has vastly lower heat capacity per gram and per volume (4000) and less than a tenth the conductivity, but also much lower viscosity (about 200 times lower: 17.4×10^{-6} Pa·s for air vs 8.94×10^{-4} Pa·s for water). Continuing the calculation from two paragraphs above, air cooling needs ten times of the surface area, therefore the fins, and air needs 2000 times the flow velocity and thus a recirculating air fan needs ten times the power of a recirculating water pump. Moving heat from the cylinder to a large surface area for air cooling can present problems such as difficulties manufacturing the shapes needed for good heat transfer and the space needed for free flow of a large volume of air. Water boils at about the same temperature desired for engine cooling. This has the advantage that it absorbs a great deal of energy with very little rise in temperature (called heat of vaporization), which is good for keeping things cool, especially for passing one stream of coolant over several hot objects and achieving uniform temperature. In contrast, passing air over several hot objects in series warms the air at each step, so the first may be over-cooled and the last under-cooled. However, once water boils, it is an insulator, leading to a sudden loss of cooling where steam bubbles form (for more, see heat transfer). Steam may return to water as it mixes with other coolant, so an engine temperature gauge can indicate an acceptable temperature even though local temperatures are high enough that damage is being done.

An engine needs different temperatures. The inlet including the compressor of a turbo and in the inlet trumpets and the inlet valves need to be as cold as possible. A countercurrent heat exchange with forced cooling air does the job. The cylinder-walls should not heat up the air before compression, but also not cool down the gas at the combustion. A compromise is a wall temperature of 90 °C. The viscosity of the oil is optimized for just this temperature. Any cooling of the exhaust and the turbine of the turbocharger reduces the amount of power available to the turbine, so the exhaust system is often insulated between engine and turbocharger to keep the exhaust gases as hot as possible.

The temperature of the cooling air may range from well below freezing to 50 °C. Further, while engines in long-haul boat or rail service may operate at a steady load, road vehicles often see widely varying and quickly varying load. Thus, the cooling system is designed to vary cooling so the engine is neither too hot nor too cold. Cooling system regulation includes adjustable baffles in the air flow (sometimes called 'shutters' and commonly run by a pneumatic 'shutterstat'); a fan which operates either independently of the engine, such as an electric fan, or which has an adjustable clutch; a thermostatic valve or just 'thermostat' that can block the coolant flow when too cool. In addition, the motor, coolant, and heat exchanger have some heat capacity which smooths out temperature increase in short sprints. Some engine controls shut down an engine or limit it to half throttle if it overheats. Modern electronic engine controls adjust cooling based on throttle to anticipate a temperature rise, and limit engine power output to compensate for finite cooling.

Finally, other concerns may dominate cooling system design. As example, air is a relatively poor coolant, but air cooling systems are simple, and failure rates typically rise as the square of the number of failure points. Also, cooling capacity is reduced only slightly by small air coolant leaks. Where reliability is of utmost importance, as in aircraft, it may be a good trade-off to give up efficiency, longevity (interval between engine rebuilds), and quietness in order to achieve slightly higher reliability; the consequences of a broken airplane engine are so severe, even a slight increase in reliability is worth giving up other good properties to achieve it.

Air-cooled and liquid-cooled engines are both used commonly. Each principle has advantages and disadvantages, and particular applications may favor one over the other. For example, most cars and trucks use liquid-cooled engines, while many small airplane and low-cost engines are air-cooled.

Generalization Difficulties

It is difficult to make generalizations about air-cooled and liquid-cooled engines. Air-cooled Deutz diesel engines are known for reliability even in extreme heat, and are often used in situations where the engine runs unattended for months at a time.

Similarly, it is usually desirable to minimize the number of heat transfer stages in order to maximize the temperature difference at each stage. However, Detroit Diesel 2-stroke cycle engines commonly use oil cooled by water, with the water in turn cooled by air.

The coolant used in many liquid-cooled engines must be renewed periodically, and can freeze at ordinary temperatures thus causing permanent engine damage. Air-cooled engines do not require coolant service, and do not suffer engine damage from freezing, two commonly cited advantages for air-cooled engines. However, coolant based on propylene glycol is liquid to -55 °C, colder than is encountered by many engines; shrinks slightly when it crystallizes, thus avoiding engine damage; and has a service life over 10,000 hours, essentially the lifetime of many engines.

It is usually more difficult to achieve either low emissions or low noise from an air-cooled engine, two more reasons most road vehicles use liquid-cooled engines. It is also often difficult to build large air-cooled engines, so nearly all air-cooled engines are under 500 kW (670 hp), whereas large liquid-cooled engines exceed 80 MW (107000 hp) (Wärtsilä-Sulzer RTA96-C 14-cylinder diesel).

Air-cooling

Cars and trucks using direct air cooling (without an intermediate liquid) were built over a long period from the very beginning and ending with a small and generally unrecognized technical change. Before World War II, water-cooled cars and trucks routinely overheated while climbing mountain roads, creating geysers of boiling cooling water. This was considered normal, and at the time, most noted mountain roads had auto repair shops to minister to overheating engines.

ACS (Auto Club Suisse) maintains historical monuments to that era on the Susten Pass where two radiator refill stations remain. These have instructions on a cast metal plaque and a spherical bottom watering can hanging next to a water spigot. The spherical bottom was intended to keep it from being set down and, therefore, be useless around the house, in spite of which it was stolen, as the picture shows.

During that period, European firms such as Magirus-Deutz built air-cooled diesel trucks, Porsche built air-cooled farm tractors, and Volkswagen became famous with air-cooled passenger cars. In the United States, Franklin built air-cooled engines.

For many years air cooling was favored for military applications as liquid cooling systems are more vulnerable to damage by shrapnel.

The Czechoslovakia based company Tatra is known for their large displacement air-cooled V8 car engines; Tatra engineer Julius Mackerle published a book on it. Air-cooled engines are better adapted to extremely cold and hot environmental weather temperatures: you can see air-cooled engines starting and running in freezing conditions that seized water-cooled engines and continue working when water-cooled ones start producing steam jets. Air-cooled engines have may be an advantage from a thermodynamic point of view due to higher operating temperature. The worst problem met in air-cooled aircraft engines was the so-called "Shock cooling", when the airplane entered in a dive after climbing or level flight with throttle open, with the engine under no load while the airplane dives generating less heat, and the flow of air that cools the engine is increased, a catastrophic engine failure may result as different parts of engine have different temperatures, and thus different thermal expansions. In such conditions, the engine may seize, and any sudden change or imbalance in the relation between heat produced by the engine and heat dissipated by cooling may result in an increased wear of engine, as a consequence also of thermal expansion differences between parts of engine, liquid-cooled engines having more stable and uniform working temperatures.

Liquid Cooling

Today, most automotive and larger IC engines are liquid-cooled.

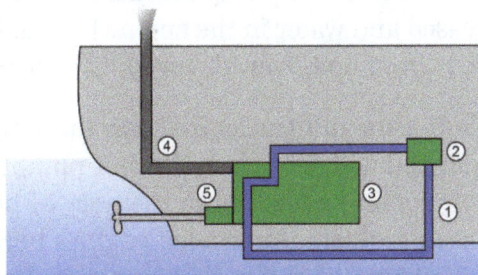

A fully closed IC engine cooling system

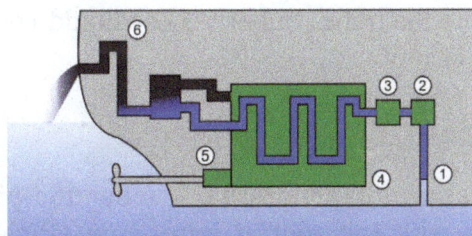

Open IC engine cooling system

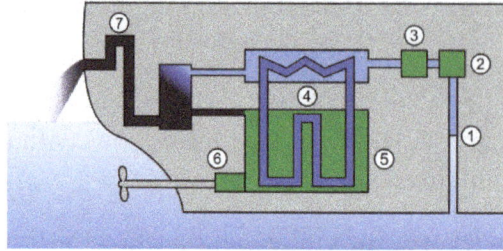

Semiclosed IC engine cooling system

Liquid cooling is also employed in maritime vehicles (vessels, ...). For vessels, the seawater itself is mostly used for cooling. In some cases, chemical coolants are also employed (in closed systems) or they are mixed with seawater cooling.

Transition from Air Cooling

The change of air cooling to liquid cooling occurred at the start of World War II when the US military needed reliable vehicles. The subject of boiling engines was addressed, researched, and a solution found. Previous radiators and engine blocks were properly designed and survived durability tests, but used water pumps with a leaky graphite-lubricated "rope" seal (gland) on the pump shaft. The seal was inherited from steam engines, where water loss is accepted, since steam engines already expend large volumes of water. Because the pump seal leaked mainly when the pump was running and the engine was hot, the water loss evaporated inconspicuously, leaving at best a small rusty trace when the engine stopped and cooled, thereby not revealing significant water loss. Automobile radiators (or heat exchangers) have an outlet that feeds cooled water to the engine and the engine has an outlet that feeds heated water to the top of the radiator. Water circulation is aided by a rotary pump that has only a slight effect, having to work over such a wide range of speeds that its impeller has only a minimal effect as a pump. While running, the leaking pump seal drained cooling water to a level where the pump could no longer return water to the top of the radiator, so water circulation ceased and water in the engine boiled. However, since water loss led to overheat and further water loss from boil-over, the original water loss was hidden.

After isolating the pump problem, cars and trucks built for the war effort (no civilian cars were built during that time) were equipped with carbon-seal water pumps that did not leak and caused no more geysers. Meanwhile, air cooling advanced in memory of boiling engines... even though boil-over was no longer a common problem. Air-cooled engines became popular throughout Europe. After the war, Volkswagen advertised in the USA as not boiling over, even though new water-cooled cars no longer boiled over, but these cars sold well. But as air quality awareness rose in the 1960s, and laws governing exhaust emissions were passed, unleaded gas replaced leaded gas and leaner fuel mixtures became the norm. Subaru chose liquid-cooling for their EA series (flat) engine when it was introduced in 1966.

Low Heat Rejection Engines

A special class of experimental prototype internal combustion piston engines have been developed over several decades with the goal of improving efficiency by reducing heat loss. These engines are variously called adiabatic engines, due to better approximation of adiabatic expansion, low heat rejection engines, or high temperature engines. They are generally diesel engines with combustion chamber parts lined with ceramic thermal barrier coatings. Some make use of titanium pistons and other titanium parts due to its low thermal conductivity and mass. Some designs are able to eliminate the use of a cooling system and associated parasitic losses altogether. Developing lubricants able to withstand the higher temperatures involved has been a major barrier to commercialization.

Consumption Map

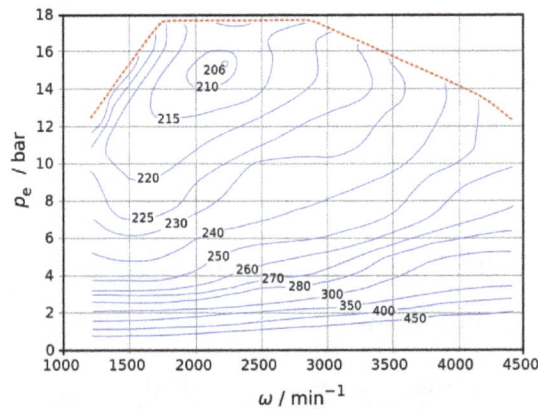

Consumption map of a three cyllinder diesel engine, capacity1.5 l

The consumption map or efficiency map shows the brake specific fuel consumption in g per kWh over mean effective pressure per rotational speed of an internal combustion engine.

On the abscissa it shows the rotational speed range. The ordinate is limited by the maximum load of the engine. The lines show the specific fuel consumption and appear similar to a shell.

The map contains each possible condition, combining rotational speed mean effective pressure. It shows the result of specific fuel consumption. A typical rotation power output P (linear to $p_e \cdot \omega$) is reached on several locations on the map but differing in the amount of fuel consumption. Automatic transmissions, are designed to keep the engine in the lowest possible fuel consumtion mode.

The map also shows the effiency of the engine. Depending on the fuel type, diesel and gasoline engines reach up to 210 g per kWh and about 40% of efficiency. Using natural gas this efficiency is reached at 200 g per kWh.

Average values are 160 to 180 g per kWh for slow moving two stroke diesel boat engines using fuel oil, reaching up to 55% efficiency at 300 rpm. 210 to 195 g per kWh at cooled and pre charged diesel engines for passenger cars, trucks 225 to 195 g per kWh. Non charged otto cycle gasoline engines for passenger cars 350 to 250 g per kWh.

Literature

- (German) Richard van Basshuysen: *Handbuch Verbrennungsmotor*, Fred Schäfer; 3. Auflage; 2005; Vieweg Verlag

Brake Specific Fuel Consumption

Brake specific fuel consumption (BSFC) is a measure of the fuel efficiency of any prime mover that burns fuel and produces rotational, or shaft, power. It is typically used for comparing the efficiency of internal combustion engines with a shaft output.

It is the rate of fuel consumption divided by the power produced. It may also be thought of as power-specific fuel consumption, for this reason. BSFC allows the fuel efficiency of different engines to be directly compared.

The BSFC Calculation (in Metric Units)

To calculate BSFC, use the formula

$$BSFC = \frac{r}{P}$$

where:

r is the fuel consumption rate in grams per second (g/s)

P is the power produced in watts where

ω is the engine speed in radians per second (rad/s)

τ is the engine torque in newton meters (N·m)

The above values of r, ω, and τ may be readily measured by instrumentation with an engine mounted in a test stand and a load applied to the running engine. The resulting units of BSFC are grams per joule (g/J)

Commonly BSFC is expressed in units of grams per kilowatt-hour (g/(kW·h)). The conversion factor is as follows:

BSFC [g/(kW·h)] = BSFC [g/J]×(3.6×10^6)

The conversion between metric and imperial units is:

BSFC [g/(kW·h)] = BSFC [lb/(hp·h)]×608.277

BSFC [lb/(hp·h)] = BSFC [g/(kW·h)]×0.001644

The Relationship Between BSFC Numbers and Efficiency

To calculate the actual efficiency of an engine requires the energy density of the fuel being used.

Different fuels have different energy densities defined by the fuel's heating value. The lower heating value (LHV) is used for internal combustion engine efficiency calculations because the heat at temperatures below 150 °C (300 °F) cannot be put to use.

Some examples of lower heating values for vehicle fuels are:

Certification gasoline = 18,640 BTU/lb (0.01204 kW·h/g)

Regular gasoline = 18,917 BTU/lb (0.0122222kW·h/g)

Diesel fuel = 18,500 BTU/lb (0.0119531 kW·h/g)

Thus a diesel engine's efficiency = 1/(BSFC × 0.0119531) and a gasoline engine's efficiency = 1/(BSFC × 0.0122225)

The Use of BSFC Numbers as Operating Values and as a Cycle Average Statistic

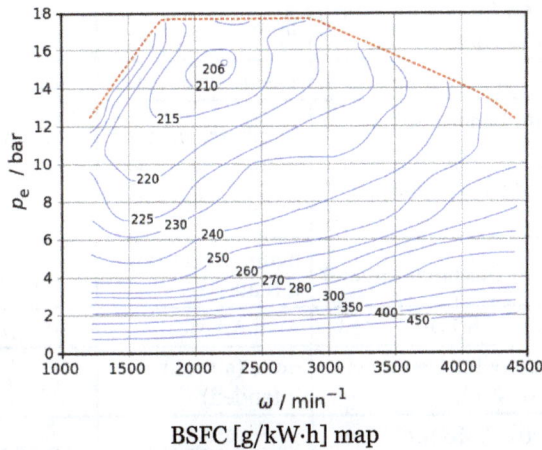

BSFC [g/kW·h] map

Any engine will have different BSFC values at different speeds and loads. For example, a reciprocating engine achieves maximum efficiency when the intake air is unthrottled and the engine is running near its peak torque. The efficiency often reported for a particular engine, however, is not its maximum efficiency but a fuel economy cycle statistical average. For example, the cycle average value of BSFC for a gasoline engine is 322 g/kW·h, translating to an efficiency of 25% (1/(322 × 0.0122225) = 0.2540). Actual efficiency can be lower or higher than the engine's average due to varying operating conditions. In the case of a production gasoline engine, the most efficient BSFC is approximately 225 g/kW·h, which is equivalent to a thermodynamic efficiency of 36%.

An iso-BSFC map (fuel island plot) of a diesel engine is shown. The sweet spot at 206 BSFC has 40.6% efficiency. The x-axis is rpm; y-axis is BMEP in bar (bmep is proportional to torque)

The Significance of BSFC Numbers for Engine Design and Class

BSFC numbers change a lot for different engine design and compression ratio and power rating. Engines of different classes like diesels and gasoline engines will have very different BSFC numbers, ranging from less than 200 g/kW·h (diesel at low speed and high torque) to more than 1,000 g/kW·h (turboprop at low power level).

Examples of Values of BSFC for Shaft Engines

The following table takes values as an example for the specific fuel consumption of several types of engines. For specific engines values can and often do differ from the table values shown below.

Energy efficiency is based on a lower heating value of 42.7 MJ/kg (84.3 g/kW·h) for diesel fuel and jet fuel, 43.9 MJ/kg (82 g/kW·h) for gasoline.

Power (kW)	Year	Engine type	Application	SFC (lb/hp·h)	SFC (g/kW·h)	Energy efficiency
2,050	1996	Pratt & Whitney Canada PW127 turboprop	ATR 72 regional airliner	0.477	290	29.1%
95	1970	Lycoming O-320 piston, gasoline	General aviation	0.460	280	29.3%
63	1991	GM Saturn I4 engine, gasoline	Saturn S-Series cars	0.411	250	32.5%
150	2011	Ford EcoBoost gasoline, turbo	Ford cars	0.403	245	33.5%
2,000	1945	Wright R-3350 Duplex-Cyclone gasoline, turbo-compound	Bombers, airliners	0.380	231	35.5%
57	2003	Toyota 1NZ-FXE, gasoline	Toyota Prius car	0.370	225	36.4%
550	1931	Junkers Jumo 204 two-stroke diesel, turbo	Bombers, airliners	0.347	211	40%
36,000	2002	Rolls-Royce Marine Trent turboshaft	Combat ships	0.340	207	40.7%
2,340	1949	Napier Nomad Diesel-compound	planned (aircraft intended)	0.340	207	40.7%
165	2000	Volkswagen 3.3 V8 TDI	Audi A8 car	0.337	205	41.1%
2,013	1940	Klöckner-Humboldt-Deutz DZ 710 Diesel two stroke	none (aircraft intended)	0.330	201	41.9%
42,428	1993	General Electric LM6000 turboshaft	Ship, electricity	0.329	200.1	42.1%
130	2007	BMW N47 2L turbodiesel	BMW cars	0.326	198	42.6%
88	1990	Audi 2.5L TDI	Audi 100 car	0.326	198	42.6%
3,600		MAN Diesel 6L32/44CR four-stroke	Ship, electricity	0.283	172	49%
34,320	1998	Wärtsilä-Sulzer RTA96-C two-stroke	Ship, electricity	0.263	160	52.7%
27,060		MAN Diesel S80ME-C9.4-TII two-stroke	Ship, electricity	0.254	154.5	54.6%

Turboprops efficiency are only good at high power, for approach at low power (30% P_{max}) and especially at idle (7% P_{max}), SFC increases dramatically :

2,050 kW Pratt & Whitney turboprop PW127 (1996)				
Mode	Power	fuel flow	SFC	Energy efficiency
Nominal idle (7%)	192 hp (143 kW)	3.06 kg/min (405 lb/h)	1,282 g/kW·h (2.108 lb/hp·h)	6.6%
Approach (30%)	825 hp (615 kW)	5.15 kg/min (681 lb/h)	502 g/kW·h (0.825 lb/hp·h)	16.8%
Max cruise (78%)	2,132 hp (1,590 kW)	8.28 kg/min (1,095 lb/h)	312 g/kW·h (0.513 lb/hp·h)	27%
Max climb (80%)	2,192 hp (1,635 kW)	8.38 kg/min (1,108 lb/h)	308 g/kW·h (0.506 lb/hp·h)	27.4%
Max contin. (90%)	2,475 hp (1,846 kW)	9.22 kg/min (1,220 lb/h)	300 g/kW·h (0.493 lb/hp·h)	28.1%

Take-off (100%)	2,750 hp (2,050 kW)	9.9 kg/min (1,310 lb/h)	290 g/kW·h (0.477 lb/hp·h)	29.1%

References

- Dieter Lohse und Werner Schnabel: Grundlagen der Straßenverkehrstechnik und der Verkehrsplanung: Band 1, Beuth Verlag, 2011, ISBN 9783410172710.

- Table 1 in Konrad Reif, "Dieselmotor-Management im Überblick", Abschnitt Springer Fachmedien Wiesbaden 2014, ISBN 978-3-658-06554-6

- "Advanced Gasoline Turbocharged Direct Injection (GTDI) Engine Development" (PDF). Ford Research and Advanced Engineering. May 13, 2011.

6

Progress of Internal Combustion Engine Over the Years

Various engineers and scientists have contributed in the development of the internal combustion engine. Samuel Brown patented the first internal combustion engine which was proceeded by the design put forward by Nikolaus Otto. The history explicated in this chapter is very essential, as it educates the reader about the progress of the internal combustion engine.

History of the Internal Combustion Engine

Prior to 1860

Early internal combustion engines were used to power farm equipment similar to these models.

- 3rd century: The earliest evidence of a crank and connecting rod mechanism dates to the 3rd century AD Hierapolis sawmill in Asia Minor (Turkey) as part of the Roman Empire.

- 5th century: Roman engineers documented several crankshaft-connecting rod machines used for their sawmills.

- 9th century: The crank appears in the mid-9th century in several of the hydraulic devices described by the Banū Mūsā brothers in their *Book of Ingenious Devices*.

- In 1206, al-Jazari invented an early crankshaft, which he incorporated with a crank-connecting rod mechanism in his twin-cylinder pump. Like the modern crankshaft, Al-Jazari's mechanism consisted of a wheel setting several crank pins into motion, with the wheel's motion being circular and the pins moving back-and-forth in a straight line. The crankshaft described by al-Jazari transforms continuous rotary motion into a linear reciprocating motion,

- 17th century: Samuel Morland experiments with using gunpowder to drive water pumps.

- 17th century: Christiaan Huygens designs gunpowder to drive water pumps, to supply 3000 cubic meters of water/day for the Versailles palace gardens, essentially creating the first idea of a rudimentary internal combustion piston engine.

- 1780s: Alessandro Volta built a toy electric pistol in which an electric spark exploded a mixture of air and hydrogen, firing a cork from the end of the gun.

- 1791: John Barber receives British patent #1833 for *A Method for Rising Inflammable Air for the Purposes of Producing Motion and Facilitating Metallurgical Operations*. In it he describes a turbine.

- 1794: Robert Street built a compressionless engine. He was also the first to use liquid fuel in an internal combustion engine.

- 1794: Thomas Mead patents a gas engine.

- 1798: John Stevens builds the first double-acting, crankshaft-using internal combustion engine.

- 1801: Philippe LeBon D'Humberstein comes up with the use of compression in a two-stroke engine.

- 1807: Nicéphore Niépce installed his "moss, coal-dust and resin" fueled Pyréolophore internal combustion engine in a boat and powered up the river Saône in France. A patent was subsequently granted by Emperor Napoleon Bonaparte on 20 July 1807.

- 1807: Swiss engineer François Isaac de Rivaz built an internal combustion engine powered by a hydrogen and oxygen mixture, and ignited by electric spark. (See 1780s: Alessandro Volta above.)

- 1823: Samuel Brown patented the first internal combustion engine to be applied industrially. It was compressionless and based on what Hardenberg calls the "Leonardo cycle", which, as the name implies, was already out of date at that time.

- 1824: French physicist Sadi Carnot established the thermodynamic theory of idealized heat engines.

- 1826 April 1: American Samuel Morey received a patent for a compressionless "Gas or Vapor Engine." This is also the first recorded example of a carburetor.

- 1833: Lemuel Wellman Wright, UK patent 6525, table-type gas engine. Double-acting gas engine, first record of water-jacketed cylinder.

- 1838: A patent was granted to William Barnett, UK Patent 7615 April 1838. According to Dugald Clerk, this was the first recorded use of in-cylinder compression.

- 1853-57: Eugenio Barsanti and Felice Matteucci invented and patented an engine using the free-piston principle in an atmospheric two cycle engine.

- 1856: in Florence at *Fonderia del Pignone* (now Nuovo Pignone, later a subsidiary of General Electric), Pietro Benini realized a working prototype of the Italian engine supplying 5 HP. In subsequent years he developed more powerful engines—with one or two pistons—which served as steady power sources, replacing steam engines.

- 1857: Eugenio Barsanti and Felice Matteucci describe the principles of the free piston engine where the vacuum after the explosion allows atmospheric pressure to deliver the power stroke (British patent No 1625).

1860–1920

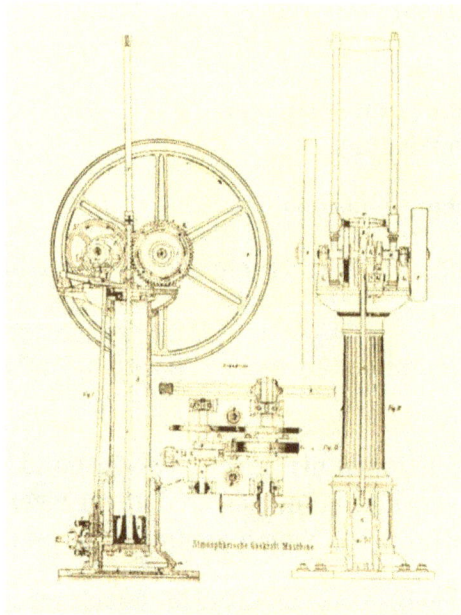

Patent of Otto-Langen engine -1863.

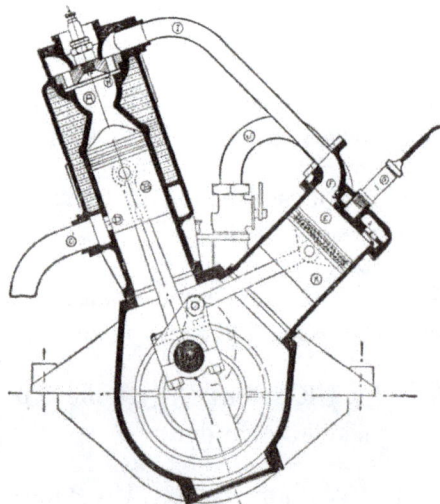

Sir Dugald Clerk's two cycle engine from 1879

This internal combustion engine was an integral aspect of the patent for the first patented automobile, made by Karl Benz on January 29, 1886

Karl Benz

- 1860: Belgian Jean Joseph Etienne Lenoir (1822–1900) produced a gas-fired internal combustion engine similar in appearance to a horizontal double-acting steam engine, with cylinders, pistons, connecting rods, and flywheel in which the gas essentially took the place of the steam. This was the first internal combustion engine to be produced in numbers.

- 1861 Nikolaus Otto builds a copy of the Lenoir engine.

- 1862 Nikolaus Otto attempts the construction of the compressed charge four cycle engine, and fails.

- 1862 The earliest confirmed patent of the 4-cycle engine, by Alphonse Beau de Rochas. This was principle only, there was NO engine built to prove the concept.

- 1862: The German Nikolaus Otto begins to manufacture a no compression gas Lenoir engine with a free piston.

- 1864: Nikolaus Otto, patented in England and other countries his first atmospheric gas engine. Otto was the first to build and sell this type of compressionless engine designed with an indirect-acting free-piston, whose great efficiency won the support of Eugen Langen and then most of the market, which at that time was mainly for small stationary engines fuelled by lighting gas. Eugen Langen collaborated with Otto in the design and they began to manufacture it in 1864.

- 1865: Pierre Hugon started production of the Hugon engine, similar to the Lenoir engine, but with better economy, and more reliable flame ignition.

- 1867: Otto and Langen exhibited their free piston engine at the Paris Exhibition in 1867, and they won the greatest award. It had less than half the gas consumption of the Lenoir or Hugon engines.

- 1870: In Vienna, Siegfried Marcus put the first mobile gasoline and the first modern Internal combustion engine on a handcart.

- 1872: In America George Brayton invented Brayton's Ready Motor and went into commercial production, this used constant pressure combustion, and was the first commercial liquid fuelled internal combustion engine.

- 1876: Nikolaus Otto, working with Gottlieb Daimler and Wilhelm Maybach, patented the compressed charge, four-cycle engine. The German courts, however, did not hold his patent to cover all in-cylinder compression engines or even the four-stroke cycle, and after this decision, in-cylinder compression became universal.

- 1878: Dugald Clerk designed the first two-stroke engine with in-cylinder compression. He patented it in England in 1881.

- 1879: Karl Benz, working independently, was granted a patent for his internal combustion engine, a reliable two-stroke gas engine. Later, Benz designed and built his own four-stroke engine that was used in his automobiles, which were developed in 1885, patented in 1886, and became the first automobiles in production.

- 1882: James Atkinson invented the Atkinson cycle engine. Atkinson's engine had one power phase per revolution together with different intake and expansion volumes, potentially making it more efficient than the Otto cycle, but certainly avoiding Otto's patent.

- 1884: British engineer Edward Butler constructed the first petrol (gasoline) internal combustion engine. Butler invented the spark plug, ignition magneto, coil ignition and spray jet carburetor, and was the first to use the word petrol.

- 1885: German engineer Gottlieb Daimler received a German patent for a supercharger

- 1887: Gustaf de Laval introduces the de Laval nozzle

- 1889: Félix Millet begins development of the first vehicle to be powered by a rotary engine in transportation history.

- 1891: Herbert Akroyd Stuart built his oil engine, leasing rights to Hornsby of England to build them. They built the first cold-start compression-ignition engines. In 1892, they installed the first ones in a water pumping station. In the same year, an experimental higher-pressure version produced self-sustaining ignition through compression alone.

- 1892: Rudolf Diesel developed the first compressed charge, compression ignition engine .

- 1893 February 23: Rudolf Diesel received a patent for his compression ignition (diesel) engine.

- 1896: Karl Benz invented the boxer engine, also known as the horizontally opposed engine, or the flat engine, in which the corresponding pistons reach top dead center at the same time, thus balancing each other in momentum.

- 1898: Fay Oliver Farwell designs the prototype of the line of Adams-Farwell automobiles, all to be powered with three or five cylinder rotary internal combustion engines.

- 1900: Rudolf Diesel demonstrated the diesel engine in the 1900 *Exposition Universelle* (World's Fair) using peanut oil fuel (see biodiesel).

- 1900: Wilhelm Maybach designed an engine built at Daimler Motoren Gesellschaft—following the specifications of Emil Jellinek—who required the engine to be named *Daimler-Mercedes* after his daughter. In 1902 automobiles with that engine were put into production by DMG.

- 1903 - Konstantin Tsiolkovsky begins a series of theoretical papers discussing the use of rocketry to reach outer space. A major point in his work is liquid fueled rockets.

- 1903: Ægidius Elling builds a gas turbine using a centrifugal compressor which runs under its own power. By most definitions, this is the first working gas turbine.

- 1905 Alfred Buchi patents the turbocharger and starts producing the first examples.

- 1903-1906: The team of Armengaud and Lemale in France build a complete gas turbine engine. It uses three separate compressors driven by a single turbine. Limits on the turbine temperatures allow for only a 3:1 compression ratio, and the turbine is not based on a Parsons-like "fan", but a Pelton wheel-like arrangement. The engine is so inefficient, at about 3% thermal efficiency, that the work is abandoned.

- 1908: New Zealand inventor Ernest Godward started a motorcycle business in Invercargill and fitted the imported bikes with his own invention – a petrol economiser. His economisers worked as well in cars as they did in motorcycles.

- 1908: Hans Holzwarth starts work on extensive research on an "explosive cycle" gas turbine, based on the Otto cycle. This design burns fuel at a constant volume and is somewhat more efficient. By 1927, when the work ended, he has reached about 13% thermal efficiency.

- 1908: René Lorin patents a design for the ramjet engine.

- 1916: Auguste Rateau suggests using exhaust-powered compressors to improve high-altitude performance, the first example of the turbocharger.

1920–1980

- 1920: William Joseph Stern reports to the Royal Air Force that there is no future for the turbine engine in aircraft. He bases his argument on the extremely low efficiency of existing compressor designs. Due to Stern's eminence, his paper is so convincing there is little official interest in gas turbine engines anywhere, although this does not last long.

- 1921: Maxime Guillaume patents the axial-flow gas turbine engine. It uses multiple stages in both the compressor and turbine, combined with a single very large combustion chamber.

- 1923: Edgar Buckingham at the United States National Bureau of Standards publishes a report on jets, coming to the same conclusion as W.J. Stern, that the turbine engine is not efficient enough. In particular he notes that a jet would use five times as much fuel as a piston engine.

- 1925: The Hesselman engine is introduced by Swedish engineer Jonas Hesselman represented the first use of direct gasoline injection on a spark-ignition engine.

- 1925: Wilhelm Pape patents a constant-volume engine design.

- 1926: Alan Arnold Griffith publishes his groundbreaking paper *Aerodynamic Theory of Turbine Design*, changing the low confidence in jet engines. In it he demonstrates that existing compressors are "flying stalled", and that major improvements can be made by re-designing the blades from a flat profile into an airfoil, going on to mathematically demonstrate that a practical engine is definitely possible and showing how to build a turboprop.

- 1926 - Robert Goddard launches the first liquid-fueled rocket

- 1927: Aurel Stodola publishes his "Steam and Gas Turbines" - basic reference for jet propulsion engineers in the USA.

- 1927: A testbed single-shaft turbo-compressor based on Griffith's blade design is tested at the Royal Aircraft Establishment.

- 1929: Frank Whittle's thesis on jet engines is published

- 1930: Schmidt patents a pulse-jet engine in Germany.

- 1935: Hans von Ohain creates plans for a turbojet engine and convinces Ernst Heinkel to develop a working model. Along with a single mechanic von Ohain develops the worlds first turbojet on a test stand.

- 1936: French engineer René Leduc, having independently re-discovered René Lorin's design, successfully demonstrates the world's first operating ramjet.

- 1937: The first successful run of Sir Frank Whittle's gas turbine for jet propulsion.

- March, 1937: The Heinkel HeS 1 experimental hydrogen fueled centrifugal jet engine is tested at Hirth.

- 27 August 1939: Flight of the world's first turbojet power aircraft. Hans von Ohain's Heinkel He 178 V1 pioneer turbojet aircraft prototype makes its first flight, powered by an He S 3 von Ohain engine.

- 15 May 1941: The Gloster E.28/39 becomes the first British jet-engined aircraft to fly, using a Power Jets W.1 turbojet designed by Frank Whittle and others.

- 1942: Max Bentele discovers in Germany that turbine blades can break if vibrations are in its resonance range, a phenomenon already known in the USA from the steam turbine experience.

- July 18, 1942: The Messerschmitt Me 262 first jet engine flight

- 1946: Samuel Baylin develops the Baylin Engine a three cycle internal combustion engine with rotary pistons. A crude but complex example of the future Wankel engine.

- 1951 engineers for The Texas Company—i.e. now Chevron—developed a four stroke engine with a fuel injector that employed what was called the Texaco Combustion Process, which unlike normal four stroke gasoline engines which used a separate valve for the intake of the air-gasoline mixture, with the T.C.P. engine the intake valve with a built in special shroud delivers the air to the cylinder in a tornado type fashion and then the fuel is injected and ignited by a spark plug. The inventors claimed their engine could burn on almost any petroleum based fuel of any octane and even some alcohol based fuels—e.g. kerosene, benzine, motor oil, tractor oil, etc. — without the pre-combustion knock and the complete burning of the fuel injected into the cylinder. While development was well advanced by 1950, there are no records of the T.C.P. engine being used commercially.

- 1950s development begins by US firms of the Free-piston engine concept which is a crankless internal combustion engine.

- 1954: Felix Wankel's first working prototype DKM 54 of the Wankel engine

1980 to Present

- 1986 Benz Gmbh files for patent protection for a form of Scotch yoke engine and begins development of same. Development subsequently abandoned.

- 1996 Ford Motor Company files patent for compact turbine engine.

- 2004 Hyper-X first scramjet to maintain altitude

- 2004 Toyota Motor Corp files for patent protection for new form of Scotch yoke engine.

Engine Starting

Early internal combustion engines were started by hand cranking. Various types of starter motor were later developed. These included:

- An auxiliary petrol engine for starting a larger petrol or diesel engine. The Hucks starter is an example

- Cartridge starters, such as the Coffman engine starter, which used a device like a blank shotgun cartridge. These were popular for aircraft engines

- Pneumatic starters

- Hydraulic starters

- Electric starters

Electric starters are now almost universal for small and medium-sized engines, while pneumatic starters are used for large engines.

Modern vs. Historical Piston Engines

The first piston engines did not have compression, but ran on an air-fuel mixture sucked or blown in during the first part of the intake stroke. The most significant distinction between modern internal combustion engines and the early designs is the use of compression of the fuel charge prior to combustion.

The problem of ignition of fuel was handled in early engines with an open flame and a sliding gate. To obtain a faster engine speed Daimler adopted a Hot Tube ignition which allowed 600 rpm immediately in his 1883 horizontal cylinder engine and very soon after over 900 rpm. Most of the engines of that time could not exceed 200 rpm due to their ignition and induction systems.

The first practical engine, Lenoir's, ran on illuminating gas (coal gas). It wasn't until 1883 that Daimler created an engine that ran on liquid petroleum, a fuel called Ligroin which has a chemical makeup of Hexane-N. The fuel is also known as petroleum naptha.

Otto's first engines were push engines which produced a push through the entire stroke (like a Diesel). Daimler's engines produced a rapid pulse, more suitable for mobile engine use.

References

- A. F. L. Beeston, M. J. L. Young, J. D. Latham, Robert Bertram Serjeant (1990), The Cambridge History of Arabic Literature, Cambridge University Press, p. 266, ISBN 0-521-32763-6

- Sally Ganchy, Sarah Gancher (2009), Islam and Science, Medicine, and Technology, The Rosen Publishing Group, p. 41, ISBN 1-4358-5066-1

- Hardenberg, Horst O. (1992). Samuel Morey and his atmospheric engine. SP-922. Warrendale, Pa.: Society of Automotive Engineers. ISBN 1-56091-240-5.

- Ricci, G.; et al. (2012). "The First Internal Combustion Engine". In Starr, Fred; et al. The Piston Engine Revolution. London: Newcomen Society. pp. 23–44. ISBN 978-0-904685-15-2.

- Zeleznik, F. J.; Mcbride, B. J. "Modeling the Internal Combustion Engine". NASA Reference Publication. NASA Technical Reports Server. Retrieved 18 October 2011.

Permissions

Index

www.ingramcontent.com/pod-product-compliance
Lightning Source LLC
Chambersburg PA
CBHW061313190326
41458CB00011B/3793